Leadership
by and for
Science Teachers

About the Author

AS A READER OF A BOOK ON SCIENCE TEACHERS' LEADERSHIP, it is fair for you to ask, "What is the author's experience?" Perhaps more important, you might ask how this experience has shaped this book. My journey as a science educator has included time as a classroom teacher and an educational leader. I had wonderful mentors from whom I learned about both excellent teaching and effective leadership.

In the late 1960s, I began teaching ninth-grade Earth science and high school biology, in addition to elementary school (K–6) science, both in the Greeley, Colorado, public schools and at the Laboratory School at the University of Northern Colorado (UNC). My first experience with excellent curriculum materials was when I used the Earth Science Curriculum Project (ESCP), Biological Sciences Curriculum Study (BSCS), and several elementary school programs—including the Science Curriculum Improvement Study (SCIS), the most helpful one. Robert B. Sund, a professor at UNC, was my mentor during this time, advising me as I pursued a master's degree and guiding me into a career in science education.

In my position at the University of Northern Colorado's Laboratory School, I had undergraduate observers in my classroom on a daily basis. I tried to represent the best possible example of science teaching. My thoughts about leadership mostly centered on my mentors and others in the science education community.

My next milestone was graduate study at New York University. Beyond the usual course work, during my second year, I worked in an elementary school in Manhattan (P.S. 33), collaborating with teachers to improve their curriculum and instruction.

My dissertation was a synthesis of science education history, philosophy of education, and educational psychology, particularly the psychology of motivation and development. The title of my dissertation—*Implications of Abraham H. Maslow's Philosophy and Psychology for Science Education in the United States*—expresses the comprehensive nature of the work. I grew immeasurably thanks to the advice of faculty such as Darrell Barnard, Morris Shamos, and Janice Gorn; the mentorship of F. J. Rutherford was surpassed by none. He brought the history of science education, an understanding of science, and experience with curriculum reform to my education.

My next career opportunity included 15 years of teaching in the education department at Carleton College, a small liberal arts college in Northfield, Minnesota. Being a faculty member opened a range of opportunities—for example, teaching a course on sustainable society with theologian Ian Barbour and broadening and deepening my understanding of educational psychology, philosophy, and science and technology policy.

In 1985, I left Carleton and joined BSCS as associate director. The next decade entailed writing proposals and assuming leadership for designing and developing innovative school science programs for elementary, middle, and high school students and undergraduate nonmajors. Under the leadership of Joseph D. McInerney, the director of BSCS, I also learned about organizational management.

In the early 1990s, I worked on national standards and chaired the science content group for this initiative at the National Research Council (NRC) in Washington, D.C. The *National Science Education Standards* (NSES) were released in late 1995 (with a 1996 copyright). Shortly before the release, I joined the NRC as executive director of the Center for Science, Mathematics, and Engineering Education (CSMEE). This role introduced me to leadership at the national level through interactions with federal agencies, congressional committees, and the executive branch. I worked closely with Gerry Wheeler, the executive director at the National Science Teaching Association (NSTA) at the time, which gave me a close perspective on science programs and teachers. Donald Kennedy, past president of Stanford University, and Bruce Alberts, then president of the National Academy of Sciences, were invaluable mentors during this work at the national level.

I returned to BSCS as executive director in 2000 and remained there until I retired in 2007, but I have continued to write books and work on *A Framework for K–12 Science Education* (NRC 2012) and the *Next Generation Science Standards* (NGSS Lead States 2013).

Throughout my career, I have published articles in refereed journals such as the *Journal of Research in Science Teaching*, *Science Education*, *International Journal of Science Education*, *American Biology Teacher*, and all of NSTA's journals.

Working with mentors taught me about facing challenges, developing standards, and providing leadership. My work broadened from its original exclusive perspective on science education to include technology, engineering, and mathematics, and I benefited from perspectives that included teaching at the elementary, middle, and high school levels, as well as at the undergraduate and graduate levels. My understanding of the history of curriculum development, national policy, and psychology and philosophy influenced my view of science education. My background has offered me broad and deep perspectives of educational leadership, and I have learned that leadership requires resolving what appear to be contradictions as the aims of science education shift with changing social and political priorities.

Rodger W. Bybee

Acknowledgments

I AM THANKFUL FOR THE UNDERSTANDING AND SUPPORT of numerous individuals. First and foremost, I am grateful to my family, all of whom witnessed firsthand my writing (and rewriting) process and shared support, advice, and appreciation for my work.

I extend my appreciation to several individuals who contributed profiles on leadership and offered friendship and support throughout the book's development. James Short from the Carnegie Corporation of New York supported the ideas for this book from the beginning. His ideas expanded my own and his challenges clarified the themes of leadership.

Harold Pratt has been a colleague for more than four decades. Across numerous discussions, my views have changed because of Harold's questions, which are always insightful and directed toward a deeper understanding of science and education.

From the beginning of this book's development, Peter McLaren from NextGen Education, Inc., took a deep and sincere interest. He expressed his ideas and thoughtful criticisms, as well as recommendations of individuals who should be included.

I am grateful to friends and colleagues for their interest, especially during the COVID-19 pandemic. Their concern was shown in questions such as "What are you working on now?" or "How is the book coming along?" I express my appreciation to Greg DeWit, Bob and Katie Pletka, Mark Salata, Kenn Heydrick, Chris Chopyak, Richard Duschl, Brett Moulding, Julienne Lee, William Banko, and Herb and Bonnie Brunkhorst.

Shortly after the pandemic-related stay-at-home measures went into effect, I began inviting individuals to prepare profiles of their reflections on leadership. Thanks to the contributors' diverse backgrounds and range of experiences, this book has the unique quality of being both by and for science teachers, as expressed in the title.

I extend my deepest appreciation to those who contributed profiles for this book. Your contributions represent the distribution of leadership in the science education community and your unique leadership qualities.

Cassie Bess	Ken Huff	Harold Pratt
James Blake	Phil Johntson	Stephen Pruitt
Bonnie J. Brunkhorst	Dora Kastel	K. Renae Pullen
Herbert K. Brunkhorst	Cynthia Long	Cynthia Rounds
Kathy DiRanna	Peter McLaren	Felicia Ryder
Arthur Eisenkraft	Tammy Wu Moriarty	James Short
Maya Garcia	Julie Olson	John Spiegel

I extend my gratitude to Kevin Anderson, Kristen Moorhead, and Zoe Evans for your constructive recommendations on early drafts of the book.

Claire Reinburg, Rachel Ledbetter, and Andrea Silen provided early advice on important features of the project. I also acknowledge NSTA leadership such as Erika Shugart, Moria Fathy Baker, Cathy Iammartino, Tricia Shelton, and Emily Brady, all of whom supported this project during a difficult period of transition at NSTA due to COVID-19.

Byllee Simon, an invaluable assistant and a wonderful friend for years, has once again contributed to the overall book and on several occasions found errors and misstatements, which she kindly and gently addressed by asking, "Are you sure you want to say this?"

Finally, Kathryn Bybee, who provided leadership in her elementary school classrooms as coordinator for the San Diego Unified School District and the San Diego County Office of Education, deserves my deepest and sincerest appreciation. She listened, gave advice, and unconditionally supported my work.

Leadership
by and for
Science Teachers

Rodger W. Bybee

nsta Press
National Science Teaching Association

ARLINGTON, VIRGINIA

Cathy Iammartino, Director of Publications and Digital Initiatives

PRINTING AND PRODUCTION
Colton Gigot, Senior Production Manager

DESIGN, PRODUCTION, AND PROJECT MANAGEMENT
KTD+ Education Group

National Science Teaching Association
Erika C. Shugart, PhD, Executive Director

1840 Wilson Blvd., Arlington, VA 22201
NSTA.org/store
For customer service inquiries, please call 800-277-5300.

Cover image: Shutterstock/Vitaly Korovin
Illustrations: iStock/RLT_Images

ISBN 978-1-68140-971-9

A catalog record of this book is available from the Library of Congress.

Contents

PART III
Some Leaders Support Science Teachers by Providing Professional Learning

PART IV
Some Leaders Support Science Teachers by Guiding Professional Organizations

PART V
Some Leaders in Science Education Have Long, Varied, and Valued Careers

PART VI
Leadership by and for Science Teachers: Guides for the Journey

Preface

LEADERSHIP BY TEACHERS OF SCIENCE is demonstrated every day in classrooms, museums, and centers for outdoor and environmental education. Unfortunately, most science teachers do not think of their role as a leader, and many educators and the public also do not recognize their leadership. Several years ago, I realized the pervasiveness of these beliefs and asked myself, "What if we perceived science teachers as leaders?" Although I have been interested in leadership for some time, an opportunity emerged to explore answers to the question, leading to this book.

The original opportunity came in 2018 in the form of a grant to the National Science Teaching Association from the Carnegie Corporation of New York. A small portion of the grant was designated as support for me to attend presentations, workshops, and institutes that provided professional learning to help educators implement the *Next Generation Science Standards*. I was supposed to prepare a series of case studies based on my visits. This opportunity to explore the combination of science teacher leadership and professional development left me excited and ready to begin work—as soon as I completed *STEM, Standards, and Strategies for High-Quality Units* (Bybee 2020).

Then, in mid-March 2020, earlier reports of the coronavirus (COVID-19) became a reality as COVID-19 spread around the world, and governments, including in the United States, took drastic measures to limit the virus's spread. Along with most Americans, I sheltered in place and began adapting to a different lifestyle.

I soon realized that school closings, virtual teaching, and myriad other issues would likely continue for a prolonged period. I recalled that in 1999, the Biological Sciences Curriculum Study had completed a module for the National Institutes of Health called *Emerging and Re-emerging Infectious Diseases* (BSCS 1999). A brief review of the module confirmed my belief that the pandemic would affect our lives for a long time, so I had time to work on this book. However, attending meetings, visiting schools, and observing teachers were all out of the plan; instead, I would incorporate profiles of science educators with leadership experience.

While it may be difficult to imagine, one positive result of the pandemic may be a new recognition of the crucial role of science, scientists, and scientists' expertise in addressing many personal and social issues. If such recognition does occur, we will need science teachers to clearly and consistently provide leadership as they represent science in our society and in their classrooms.

I hope this book sends a positive message about science teachers' leadership to the larger science education community. Although the majority of science teachers will not become department chairs, district coordinators, state supervisors, or presidents of national organizations, there is a message to all science teachers in this book nonetheless: You are already leaders in your classrooms, and your leadership is fundamental for your students' learning.

Rodger W. Bybee
Golden, Colorado

Introduction

Why Is Leadership by and for Science Teachers Especially Important Now?

ALMOST DAILY, there are reports that support an observation that science education is in a period of significant reform. The importance of leadership by science teachers seems clear and, to many, compelling. The publication of *A Framework for K–12 Science Education* (NRC 2012) and the *Next Generation Science Standards* (NGSS Lead States 2013) has had a direct influence on states' new science education standards, but after nearly a decade, the science education community has yet to implement significant reforms of school programs and classroom practices. In addition, there are connections between these changes based on standards and the challenges related to science education based on the world's lengthy fight against the coronavirus (COVID-19) and the global consequences of climate change. These factors alone provide an answer to the question posed in the introduction title, and they form a foundation for a national mission that is larger and more important than the Sputnik satellite launch that contributed to science education reform in the late 1950s and 1960s.

The science education community's search for solutions to the complicated problems related to reform underscores the need for leadership. Numerous reports document the many reasons reform is needed and the directions this reform can take. There are understandable and acceptable solutions to many problems, but science education needs widespread leadership to achieve real and necessary reform. Leadership from science teachers and contributions from those providing professional learning for teachers are essential components to achieving reform. Science teachers have the greatest burden and heaviest responsibility for reform of their school programs and classroom practices. I say this with a recognition of teachers' essential position in contemporary reform and the need for change, as well as with compassion for their difficult task.

With the COVID-19 pandemic came schools closing and adults tasked with homeschooling their children, so educators rose to the occasion and implemented distance learning. The education system and adults caring for children adapted to the extraordinary situation.

Students eventually began going back to school and many parents went back to work. But things had changed. How much students learned during school closures will be difficult for educational researchers to ascertain. After the pandemic, more of the public has a greater appreciation for both the complex task of teaching and the complicated public school system. For many, the pandemic resulted in a deeper respect for education in general and teaching in particular.

As if the pandemic was not enough to show the need for leadership, we must recognize the increasingly dire consequences of climate change. Severe weather events, extreme wildfires, longer hurricane seasons, rising sea levels, and dramatic decreases in biodiversity all require innovative solutions from the science, technology, engineering, mathematics, and health communities—and leadership by and for science teachers.

Changing Perceptions of Science Teacher Leadership

Mention leadership in almost any conversation, and responses will refer to political, military, or religious leaders, both good and bad. Seldom will you hear references to educators, and if you do, they are usually references to administrators, not teachers.

But teachers' leadership matters every hour of every school day. Teachers lead students toward a better understanding of the world through traditional means such as reading, writing, mathematics, and—of particular interest in this discussion—science. What science teachers know, value, do, and are sensible about, as they educate, affects students' lives beyond that grade or course and much deeper than one lesson or activity can. Yet many individuals, including science teachers, do not view teachers as leaders.

In my career as a science teacher, teacher educator, policy maker, and adviser to local, state, national, and international programs and organizations, I have never encountered a science teacher who came to work without the explicit intention of helping students learn. Granted, some science teachers do a better job than others, but they all have good intentions.

While working on *STEM Education Now More Than Ever* (Bybee 2018) and *STEM, Standards, and Strategies for High-Quality Units* (Bybee 2020), I had a number of interactions with colleagues who challenged my emerging perceptions of science teachers as leaders. A bit of context is important. In both of those books, I suggested the need for teachers to work on STEM-based instructional materials. When I mentioned this to colleagues, they would tell me the idea would not work because teachers did not have the required knowledge and skills. Having led numerous curriculum development projects at Biological Sciences Curriculum Study (BSCS), I knew there was some merit to this view. But I persisted because science teachers for all grades find and modify lessons, construct units, and implement new science programs.

I came to realize this perspective was quite pervasive. Within the larger education community, there is a subtle and dominant set of perceptions that teachers need to improve their knowledge and abilities. This view begins in undergraduate teacher preparation and continues in many graduate programs and professional development activities. The perspective I just described is reinforced by teachers when they ask, for example, "Where are the instructional materials?" after a presentation on new state standards. Such questions do not convey leadership, but instead communicate a need for leadership—by others.

My colleague Stephen Pruitt has, on numerous occasions, wisely counseled audiences of science teachers to never express their professional work as "I am *just* a teacher." The self-deprecation expressed by *just* reveals a widely held set of perceptions by and about teachers. Use of this word conveys a view of being merely a teacher rather than being honorable or righteous; it certainly does not convey the idea of leadership.

As a result, science teachers are seen as individuals who must improve, as individuals we need to help become better rather than must work alongside. This view is reinforced in more subtle ways with sayings such as "Those who can, do; those who cannot, teach." I am not naive to the observation that science teachers are like other professionals and need updates and new information to improve their practices, so professional learning should be an integral component of every science teacher's career.

I admit to holding the view of science teachers just described. Then came the day I was working to address the need for instructional materials and curriculum reform in light of the *NGSS* and new state standards. I asked myself, "How can I think differently about the need for innovative instructional materials?" Instead of placing the burden of curriculum reform on the federal government and commercial publishers, what if we worked with teachers to improve their school science programs? Teachers will still need additional knowledge of science concepts and processes, how students learn, and what goes into the design and development of high-quality instructional materials, but it is not a great leap to the question I asked next: "What if we perceived science teachers as educational leaders?" Asking these questions, especially the latter, initiated a quest that resulted in this book.

A TEACHER'S VIEWS ON LEADERSHIP

Linda Rost—a high school science teacher in Baker, Montana—was one of four nominees for the 2020 Teacher of the Year who participated in a roundtable discussion for *Education Week* (Will 2020). When asked if she was optimistic about the direction the teaching profession was headed, Rost responded, "I think the state of the teaching profession is changing; it's evolving in a good way. We're seeing teachers evolve in their own self-perception and evolve into leaders" (Will 2020, p. 5). The interview continued with a question about Rost's platform as a national spokesperson for teachers, to which she responded, "I'm passionate about . . . providing opportunities for teachers to step into leadership positions. I think we have a lot of teachers who have a lot of skills, but they don't have an opportunity to exercise those and showcase them" (Will 2020, p. 5). These quotations highlight the ideas of someone who is clearly a science teacher leader, sees the changing nature of the profession, and provides opportunities for teachers to step into leadership positions.

Do We Need Teacher Leadership?

Recognition of teacher leadership has been with us for a long time, though often in the background. Teacher leadership was brought to the foreground, for example, in the late 1980s and 1990s through the writings of Judith Warren Little (1988) and Mark A. Smylie (1997). During this period, however, recognition of leadership by science teachers was overshadowed by standards-based reforms initiated by Project 2061's *Science for All Americans* (Rutherford and Ahlgren 1989) and the National Research Council's (1996) *National Science Education Standards*.

In the early 21st century, teacher leadership again gained recognition in educational research (see, for example, York-Barr and Duke 2004; Wenner and Campbell 2017). One could argue that teacher leadership has once again moved to the background due to new state standards, the COVID-19 pandemic, and climate change. I take a different view: Because of state and national priorities of literacy, the environment, resources, and health, we have a clear and compelling *need* to recognize teachers as a critical component of distributed leadership for education reform in STEM disciplines.

Much of the literature on teacher leadership highlights all teachers' knowledge and skills. While their knowledge and skills are important to address, we should also focus on innovations for science teaching and ways to advance our understanding of how students learn.

The profiles in this book include discussions by science teachers in—and educators beyond—the K–12 classroom. These individuals' experiences enable them to provide knowledge and skills unique to science and policies, such as state standards for science education.

What Are the Prospects for Science Teacher Leadership?

"Very good to excellent" is my short answer when asked about the prospects for science teacher leadership. For example, new science standards require a curriculum reform and complementary professional learning. The growth of professional learning communities has resulted in teachers assuming responsibilities for leadership, which many administrators have fortunately asked for, accepted, and supported.

The teaching profession faces many challenges: low salaries, little public support, and evolving demands due to the COVID-19 pandemic, to name a few. That said, I have to view the prospects for science teacher leadership with optimism, for several reasons. Here I acknowledge the influence of Barnett Berry (2019).

First, the public now understands teachers' essential role. One result of the various state mandates to control the spread of COVID-19 could well be an understanding of and respect for teachers and the job entrusted to them. The fact that school systems, and especially teachers, adapted to the requirements for distance learning and continued to be "on call" during such a challenging period should bolster support for science teacher leadership.

Second, the effects of COVID-19 brought a new and deep appreciation for science, especially as it relates to public health. Relying on data and evidence rather than ideology and politics has been essential in efforts to mitigate the pandemic's impact.

Third, there is increasing evidence of the benefits of teacher leadership in general (see, for example, Berry 2019; Will 2017). Ingersoll and his colleagues demonstrated a correlation (not causation) between schools with high levels of instructional leadership from teachers and high scores on mathematics and English language arts state assessments compared with schools that have lower levels of teacher leadership.

Fourth, the knowledge base about teacher leadership continues to grow. The thorough review by York-Barr and Duke (2004) and a subsequent review by Wenner and Campbell (2017) both provided a theoretical and empirical basis for teacher leadership.

Finally, national groups have recognized teacher leaders. The National Science Teaching Association has an ambassador program, as does the National Center for Science Education; these programs introduce social trends, political issues, and educational priorities to emerging science education leaders. Other organizations supporting science teacher leadership include TERC and the National Academies of Science, Engineering, and Medicine.

All of these reasons—and the fact that many science teachers are already in leadership roles—inform my optimism for the future of science teachers' leadership.

An Example of Science Teacher Leadership

What roles do science teacher leaders demonstrate? What does the leadership entail? As science teachers grow professionally, they may respond to situations such as new science standards by considering questions such as "What are the standards?" and "What do the standards mean for my curriculum, my teaching, and especially my students?" But assuming the role of leadership beyond the classroom, particularly when one still has responsibilities as a teacher or will most likely return to the classroom, presents different circumstances.

In an effort to clarify the work of science teacher leaders, a research team led by Rebecca Cheung worked with a large urban school district that was anticipating the instructional demands of implementing the *NGSS* (NGSS Lead States 2013). Each school in the district designated one full-time teacher as a science teacher leader to help facilitate successful implementation. The research team aimed to support the teacher leaders and record their ideas based on interviews, focus group discussions, and written reflections. The research team subsequently organized the ideas into a framework that describes the specific tasks and work within the broader category of teacher leadership (Cheung et al. 2018). The research resulted in a science teacher leader profile that divides the role of teacher leadership into four categories: advocating, collaborating, modeling, and providing resources. Each category has two elements: one describing the work individual leaders do to further their knowledge and skills, and the other clarifying the work they do with other school personnel.

Science teacher leaders' use of this example's language and categories resulted in them having a sense of ownership, a basis for professional development, and a capacity for continued reflection on their roles. As individuals consider leadership, these roles—advocating, collaborating, modeling, and providing resources—are concrete ones to consider, as they extend one's leadership beyond the classroom to the larger science education community.

A Preview of This Book

The title of this book, *Leadership by and for Science Teachers*, suggests two complementary topics: leadership by science teachers and leadership by those who provide professional learning that enhances teachers' knowledge, skills, and ability to lead.

While developing this book, I wanted to provide science teachers with positive perceptions of themselves, their potential leadership, and their contributions to society. I also wanted to show support for science teachers as both learners and leaders. To address these goals, the book includes 21 profiles of leadership from science educators in a variety of positions and career stages. The stories describe different pathways to leadership, roles one may assume as a leader, and recommendations for effective leadership.

The next figures include brief previews of the book's parts, chapters, and profiles by contributors. There is a balance of personal stories by leaders, discussions of knowledge about teacher leadership, and insights and recommendations for those embarking on a personal journey of leadership.

PART I. Most Leadership in Science Education Is by Classroom Teachers

CHAPTER 1 *My Science Teaching*	CHAPTER 2 *My Classroom and a Vision of Where the Rubber Meets the Road*	CHAPTER 3 *Leadership as an Early-Career Elementary School Teacher*	CHAPTER 4 *Achieving Effectiveness as a Teacher Leader*	CHAPTER 5 *From The Classroom to Leadership and Back*
Julie A. Olson High School Science Teacher *Mitchell High School Mitchell, SD*	**Kenneth L. Huff** Middle School Science Teacher *Williamsville Central School District Williamsville, NY*	**Cassie Bess** Elementary School Teacher *Solana Beach School District Solana Beach, CA*	**Cynthia Rounds** Middle School Science Teacher *Fullerton, CA*	**Felicia Ryder** Middle School Science Teacher *Wangenheim Middle School San Diego, CA*
• Trust in yourself. • Recognize the worth of those you are leading. • Model resilience.	• Have a vision. • Be aware of your knowledge and abilities. • Empower others.	• Have dedication. • Self-reflect. • Form connections.	• Be bold and prepared. • Review where you are and want to be. • Seek advice from mentors.	• Collaborate. • Build connections.

PART II. Some Leaders Leave the Classroom to Support Science Teachers

CHAPTER 6	CHAPTER 7	CHAPTER 8	CHAPTER 9	CHAPTER 10
From a Science Teacher to Educational Leadership: Recognizing Common and Uncommon Opportunities	*Being a Teacher Leader*	*My Leadership Journey*	*Leadership for STEM Education*	*My Path to Education Leadership*
James Blake Director of Strategic Partnerships and Focus Programs, Science Focus Program Principal, Arts and Humanities Focus Program *Lincoln Public Schools Lincoln, NE*	**K. Renae Pullen** Elementary School Curriculum and Instructional Specialist *Caddo Parish Public Schools Shreveport, LA*	**Cynthia Long** Science Curriculum Coordinator, K–12 *School District of Osceola County Kissimmee, FL*	**Philip Johnston** STEM Coordinator *Central High School Grand Junction, CO*	**Maya M. Garcia** Supervisor and Chief Program Officer *Beyond100K Denver, CO*
• Embrace change. • Seek mentors. • Collaborate with colleagues.	• Care for others. • Be brave. • Learn from other leaders.	• Be a learner. • Do your homework. • Recognize different perspectives.	• Value personal relationships. • Make learning meaningful. • Have a vision.	• Begin with a clear vision and strategic plan. • Listen to mentors. • Do not lose sight of equity.

PART III. Some Leaders Support Science Teachers by Providing Professional Learning

CHAPTER 11	CHAPTER 12	CHAPTER 13	CHAPTER 14
Challenges and Leadership in Science Teacher Education **Herbert K. Brunkhorst** Science Education Professor (Emeritus) *California State University, San Bernardino* *San Bernardino, CA*	*Teacher Leadership: Personal Experiences, School Partnerships, and Professional Perspectives* **Tammy Wu Moriarty** Associate Director *Center to Support Excellence in Teaching* *Stanford University* *Stanford, CA*	*Leading Curriculum Reform* **Dora Kastel** Director of Curriculum Instruction (Content Area) *New Visions for Public Schools* *New York, NY*	*Six Stories From a Leadership Journey* **John P. Spiegel** Director of Curriculum and Instruction *San Diego County Office of Education* *San Diego, CA*
• Put outcomes for students first. • Pay attention to personal relations. • Don't stop learning.	• Value relationships. • Have a clear plan. • Emphasize instructional practices.	• Before anything, set goals. • Find opportunities for relationship building. • Align professional learning with instructional materials.	• Find a passion. • Serve others. • Have confidence.

PART IV. Some Leaders Support Science Teachers by Guiding Professional Organizations

CHAPTER 15	CHAPTER 16	CHAPTER 17
Beyond the Classroom: Serving All Students **Stephen L. Pruitt** President *Southern Regional Education Board* *Atlanta, GA*	*Led to Lead* **Peter J. McLaren** Executive Director *Next Gen Education, LLC* *North Kingston, RI*	*Learning to Lead Curriculum Implementation* **James B. Short** Program Director *Carnegie Corporation of New York* *New York, NY*
• Emphasize equity. • Value the individuals you are leading. • Have courage and patience.	• Leadership requires courage. • Leadership requires perspective.	• Recognize connections between leadership and learning. • Pay attention to curriculum. • Rethink professional learning.

PART V. Some Leaders in Science Education Have Long, Varied, and Valued Careers

CHAPTER 18 *Developing Leadership in Science Education*	CHAPTER 19 *Fulfilling a Legacy of Leadership in Science Education*	CHAPTER 20 *A Passion for Science and Teaching*	CHAPTER 21 *My Evolution as a Teacher Leader*
Kathy DiRanna Director (Emeritus) *K–12 Science Alliance at WestEd San Francisco, CA*	**Bonnie J. Brunkhorst** *Professor (Emeritus), Geological Sciences and Science Education California State University, San Bernardino San Bernardino, CA*	**Arthur Eisenkraft** *Distinguished Professor of Science Education, Professor of Physics, and Director of the Center of Science and Math in Context (COSMIC) University of Massachusetts Boston Boston, MA*	**Harold A. Pratt** *Director of Science (Retired) Jefferson County Public Schools Golden, CO*
• Take responsibility for something you care about. • Leadership is about people. • Pay attention to the politics.	• Recognize opportunities for leadership. • Listen to and learn from mentors. • Show understanding and respect for those affected by your leadership.	• Speak up for yourself and others. • Overcome fear and anxiety. • Embrace the challenge.	• Focus on leading, not being a leader. • Identify a mentor and listen to him/her. • Be authentically enthusiastic.

This final chapters in the book (Part VI) offer my synthesis of ideas, recommendations from the contributors, and a brief conclusion. The profiles provide a variety of insights about leadership, difficulties leaders face, and contributions leaders have made to science education. I have also done my best to synthesize the many reflections on leadership. The synthesis was outlined using words that begin with the eight letters in PROFILES: Purpose, Relationships, Opportunities, Facilitation, Integrity, Learner, Equity, and Systems. I propose these words as guides for leaders, whether they are in classrooms, state agencies, colleges or universities, or professional organizations.

Conclusion

This book's title signals the relationship between science teachers' leadership and the professional learning provided by other leaders. Although science teachers, like any professionals, require opportunities to learn, they also need to recognize the possibilities of their role as leaders. Even more important, the larger science education community has an obligation to acknowledge science teachers as leaders.

Leadership in science education is distributed across the classroom, school, state, and national levels and through a variety of individuals and institutions, with this distribution ensuring the vitality of such a large-scale system. It is impossible for one person or organization to lead all science education reform, but everyone from individuals to national organizations can assume leadership for some responsibilities. With states adopting innovative standards for science education, students returning to school after distance learning, and challenges we face as a country and the world at large, the answer to the question in this chapter's title could not be clearer.

The COVID-19 pandemic has made clear the essential role of both science teachers and science as a field for addressing the challenges we face. Science teachers' leadership is more important than ever.

PART I

Most Leadership in Science Education Is by Classroom Teachers

CHAPTER

1

My Science Teaching

JULIE A. OLSON

Julie A. Olson is a high school science teacher at Mitchell High School in Mitchell, South Dakota.

NOTE FROM RODGER

The volleyball game began at 7 p.m. and Julie Olson was announcing the game. Before the game began, there was a teacher appreciation night for Mitchell High School faculty to show gratitude for how teachers lead students' academic lives and extracurricular activities.

I met and worked with Julie on the development of the *Next Generation Science Standards*. Julie has a bachelor of science degree in biology with a minor in chemistry, as well as a master's degree in biology, both from the University of South Dakota. In the late 1980s, Julie began teaching high school science and math in rural South Dakota. The majority of her professional activities have contributed to her expertise as a classroom teacher.

The following profile of Julie includes a portion that was first published in *The Science Teacher* (Olson 2019). She has adapted and updated it for inclusion in this book.

My Teaching

A small, rural school (750 students in grades 9–12), Mitchell High School offers a variety of common classes and dual-credit, online, and career and technical education (CTE) opportunities. We also have an alternative school that provides a self-paced, customized program for at-risk students.

I teach science for Second Chance HS (our alternative school). I have between 10 and 18 students at a time, each on a different course in a different spot. This situation might seem chaotic, but the one-on-one interactions, with small units making up the courses, allow me to discuss, explain, and problem solve in a short amount of time.

The required courses (physical science, biology, and chemistry) use the online Edgenuity platform, though I have chosen hands-on activities for each course. I developed elective courses in basic electronics, robotics, drones, forensics, applied biochemistry (drug, nicotine, and alcohol education), and anatomy to introduce career opportunities and spark students' interest. Our school also offers environmental science as a blended online and hands-on course.

About twice each month, we conduct classwide experiments and projects such as testing the water quality of the local lake, making cardboard box pizza ovens, studying the chemistry of pottery and glazes, and even using basic computer-assisted design to do three-dimensional (3-D) printing or laser cutting and engraving. We also have partnered with the school's culinary teacher to build and maintain a hydroponic system.

Thanks to flexibility in students' schedules, a small staff, and a central location for the approximately 80 students in the program, it is easy to collaborate with other teachers as well as engage many students in larger projects (e.g., stocking our classroom-raised trout through the Trout in the Classroom project) or community service projects (e.g., planting in the community garden, testing water quality).

I can also tailor electives and special projects to students' individual likes and needs more easily than I could in a larger standard class. For example, a physical science student who was also a very good artist built replicas of Leonardo da Vinci's inventions and studied Alexander Calder and his kinetic mobiles. The student then built a mobile based on his own hobbies and interests.

I also teach Bio-151, a dual-credit course with the Northern State University Rising Scholars Program that enables students to earn college credit. This course covers required biology concepts for the college course, as well as anatomy and student research opportunities. This is a "traditional" class in which all students have lectures and group labs and activities at the same pace.

Offering students a variety of opportunities and exposing them to a range of biology topics is one of my teaching goals. Students begin the course by researching and testing common plants for antimicrobial properties. When studying muscle physiology, students test human-to-human interface systems and get to "control" each other's muscles and movements, then engineer assistance devices for individuals with impairments. In the spring, students studying biomechanics and computers come together to create cyborg cockroaches. The advantage of this style of class is the ability to set up complex labs and research. When certain lab activities are set up, I can also offer them to the alternative school students.

After school and on weekends, I mentor students for the science and engineering fair or help Second Chance students lead Science Saturday activities for elementary students. On weeknights, I do announcing for the high school volleyball team or teach a secondary biology methods course for Dakota Wesleyan University.

Resources for Ideas and Inspiration

The communication and conferences offered by professional associations provide lots of ideas. I attend workshops and our annual state science conference, and I am a member of both the National Science Teaching Association (NSTA) and the National Association of Biology Teachers. *The Science Teacher* (*TST*) and *NSTA Reports* list many opportunities for professional development; these publications offer many ideas, so the ability to look at back issues is helpful for my own development. For instance, I wanted to incorporate engineering into a cardiovascular unit, so I modified an activity from a back issue of *TST*, as I do with many older activities to align them with the *Next Generation Science Standards* (*NGSS*). As the co-editor of the South Dakota Science Teaching Association, I am always looking for professional development opportunities to share. I have found that applying for professional development and committees and signing up for e-mail lists lead to getting more announcements and opportunities. The Second Chance and CTE staff also collaborate frequently during lunch.

One cannot be afraid to apply for committees or programs such as the National Council of Teachers of Mathematics/NSTA STEM Ambassador program or Innovation Collaborative. I was fortunate to serve on the writing teams for both the *NGSS* and South Dakota's science standards, and getting to work with other committed individuals is inspiring. Through these opportunities, I have not only learned how to promote STEM and CTE but also exchanged ideas and collaborated with creative individuals.

How to Engage Students and Encourage Innovative Thinking

I love doing hands-on activities that give students a choice of materials, procedure, or design. We conduct some prescribed labs, but the goal for these is to develop skills and knowledge that students will be able to adapt and use for open-ended activities later in the unit. Some of the best lessons are those in which students start to ask questions such as "Which is better? To dip a chip in salsa, then flip it over to dip again, or to double dip?" (This is an example from a discussion about bacteria.) As another example, when we start learning about the musculoskeletal system, I say that I cannot do push-ups because my arms are too long. Is there evidence they can gather to prove me right or wrong?

I conclude the year with a two-week "crime scene" activity. I set up a scene with evidence that all teams can sketch, tag, and photograph. Students analyze, ask questions (interrogations), look for patterns, come to conclusions, and write a scenario based on their evidence.

I have the alternative school students try 3-D printing and laser cutting, preferably using their own designs (Bio-151 students are always welcome to come in as well to investigate further and use these machines), but I have found that students need to see the capabilities first. Elective courses are designed to accommodate many career and personal interests.

Supporting and Guiding Students

When students see a purpose and have the necessary skills to pursue it, my job is to offer a problem to solve or a question to answer; make students aware of safety and ethical concerns; and provide necessary materials (or at least some of them). If students run into roadblocks, sometimes it is as easy as pointing them to a source or asking a simple question—or brainstorming with them while letting them know I have confidence in their abilities. Teachers should also realize there is merit to some projects that divert from the intended path; in those situations, my job is to support and guide students as needed.

The Most Important Big-Picture Takeaway

Science applies to everything students do. That is the important takeaway for all students. I want them to learn how to search for and analyze evidence and make sound decisions based on evidence—whether in a lab, the grocery store, or their homes, and especially as citizens.

Reflections on Leadership

Parental Influence

My parents were great role models. My mother (a former head surgical nurse) coordinated the school carnival committee for many years and knows how to plan and to trust others to do their jobs—including us kids, who were expected to help by taking tickets, running the fishing booth, and even dressing as a clown. She also led our Girl Scout troops for many years. Being the second of six kids, I was all too happy to go to Girl Scout camp. I think this was one of my first leadership roles because I was not scared to be away from home or sleep in a tent, so the other campers looked up to me. Several years later, I became a junior leader and helped plan and carry out activities for younger Girl Scouts. I still volunteer to help lead activities for the local Boy and Girl Scout troops.

My father was active in the Jaycees organization and helped build our city's first baseball stadium and tennis courts, with my siblings and I tagging along to help. We learned how to give back and lead through our parents' example. They always did their best, and I wanted to make them proud by always doing my best and giving back.

Early Experiences

In high school, I was an elementary basketball coach and co-captain of the varsity basketball team. When I was in graduate school, I led first-year biology labs, but the event that stands out from that time is when I was asked to tutor a blind biology major. One cannot simply say, "See the …" I had to use my creativity to make cut-outs and change what I was saying to accommodate this student. This was also when I started to consider teaching as a profession.

Being a Science Teacher Leader

My first teaching job was teaching biology and physiology (for one semester) for a professor taking sabbatical at the local university. There were so many things I had never done before, but I was the teacher, so I had to show that I *could* do them—such as pithing a frog!

I applied to several places and got a teaching job in rural South Dakota as the science and math teacher for the entire high school. I always refer to this experience as "trial by fire" since I hadn't student taught in math or any of the science courses aside from biology. My mother will be the first one to say how stubborn I am and that the surest way to get me to do something is to say it can't be done. Having to prepare for six different courses wasn't going to stop me. I will always remember the first lab I conducted with the chemistry class. I said, "Go get some beakers and goggles." Well, the class just stood there. When I asked what was wrong, a student said, "We don't know what those are!"

Leadership occurs at many different levels, and leaders have to trust in others and their abilities to carry out their jobs and assignments. I have to trust that my students will do what is necessary to succeed, knowing that I have properly prepared them to take those steps. As a former department chair, I had to trust my ability to lead not by controlling faculty but collaborating and sharing ideas and techniques.

Taking on positions of leadership can be frightening! I can attest that I felt trepidation every time I signed up to be on a committee, spoke at an event, or took on a new position, but I have never regretted any of those times—even when I took over for my former mentor when he retired, the most worrisome task because I looked up to him so much. He was a great leader, so I knew I had good training.

I was asked many times to run for president of our state association but kept saying I needed to be available for my children and would run when they finished high school. I didn't fathom that they would soon grow up and I'd have to keep my word! I then encouraged others to do the same. As a team, those of us at the association worked hard to bring young teachers into the organization and give them opportunities to develop leadership skills so they could become our new leaders.

There are so many opportunities for new and young teachers to develop their leadership skills. They must not be afraid to fail. Every opportunity is a chance to grow and learn—even the "failed" applications for fellowship, grants, and committees. As a former principal told me, "You might not be successful every time, but if you are doing it with the best of intentions and in the best interests of your students, you will be okay. Just regroup and continue on." As a leader, you can model resilience, give back, and have passion for your job. Geek out!

My current foray is to learn more about global education and cross-curricular, collaborative, problem-based learning. I was fortunate to be chosen for the Fulbright Education for Sustainable Development program, which involved travel to Japan and an opportunity to collaborate with teachers from all over the United States and Japan. Two years later, we are still collaborating on topics such as water quality, hydroponics, plastics pollution, and agricultural practices.

I am also pursuing a doctorate in science curriculum and instruction and hope to lead professional development for all levels and integrate all subjects when I can.

RODGER'S INSIGHTS AND INTERPRETATIONS

Based on my time in Julie's classroom and my work with her on the *NGSS*, she is an excellent example of how a midcareer science teacher can lead. Julie's profile revealed several insights about leadership by and for science teachers:

1. Leaders have to trust in others.
2. Assuming leadership can be frightening and takes courage.
3. Leaders have to model resilience.
4. Leaders should continue to expand their experiences and bring what they have learned back to the classroom.

Julie continues as a classroom teacher but also leads the way in her school, at the state level, and in national activities.

CHAPTER

2

My Classroom and a Vision of Where the Rubber Meets the Road

KENNETH L. HUFF

Kenneth L. Huff is a middle school Earth science teacher in the Williamsville Central School District in Williamsville, New York.

NOTE FROM RODGER

I met Kenneth Huff at a gathering of writing teams for the *Next Generation Science Standards* (NGSS Lead States 2013). Across several discussions with Kenneth and observations of his interactions with colleagues, I saw his clear dedication as a classroom teacher.

Kenneth is a National Board Certified teacher with more than 23 years of classroom experience. He serves as a member of his district's Staff Development Council and leads a Young Astronaut Council for fifth- through eighth-grade students.

A native of New York, Kenneth earned his bachelor's of science and master's of science degrees in education at the State University of New York at Buffalo.

My Experience

My curriculum has been inspired by my passion for Earth and space science, a passion that translates well to my students. One example is the Young Astronaut Council at Mill Middle School, which I founded and have been coordinating along with the Williamsville Space Lab Planetarium to enhance students' classroom experiences. I work with the NASA Johnson Space Center to obtain lunar samples for students to use in classroom activities and even coordinated downlink events with astronauts in which students discussed questions about space exploration, technology, and the International Space Station. One student said, "There never was a class we didn't learn something."

Beyond the classroom, I have taken an active role in education in New York State as a whole. As a member of the New York State Education Department Science Education Steering Committee and the Science Content Advisory Panel, I provide guidance on the implementation of the NYS P–12 Science Learning Standards. In 2019, I was elected president of the Science Teachers Association of New York State (STANYS) and became a STANYS fellow.

I have served as president of the Association of Presidential Awardees in Science Teaching, as co-chair of the National Academies of Sciences Teacher Advisory Council, and on the board of directors for the National Science Teaching Association (NSTA) and have been active in the National Science Education Leadership Association. I have contributed to multiple publications for the National Academies and NSTA periodicals *Science Scope* and *The Science Teacher*.

I have been honored to receive a number of awards that recognize my dedication as a science teacher, including the Presidential Award for Excellence in Mathematics and Science Teaching in 2006 and NSTA's Robert E. Yager Foundation Excellence in Teaching Award in 2015.

Reflections on Leadership

As president of STANYS, I created a vision that was captured in the title of my speech at the 2020 STANYS conference, "Transforming Innovations Into Reality in Science." The acronym is TIRS—where the rubber hits the road, specifically in our science classrooms.

Let me begin by clarifying the terms:

- *Transformation* represents new ways of thinking about science teaching and student learning.

- *Innovations* refers to how we reinvent ourselves by looking for opportunities to grow as science teachers and educators.

- *Reality* is the understanding that new standards are here, and we have an obligation as professionals to move beyond the status quo and implement science programs and practices aligned with these standards.

- *Science* includes our current understanding of natural systems and the processes by which knowledge is revised as new evidence emerges.

A majority of states are somewhere on a continuum from initial awareness and building capacity to a full reform of curriculum, instruction, and assessments aligned with new science standards. I present the TIRS vision at this time because the beneficiaries of our efforts to implement the TIRS vision will be our students—and thus the future of society.

What if we in the science education community could empower all teachers of science to transform from using traditional ideas on one end of the continuum to the new vision offered at the other end? How would this transformation improve the general state of science education? The power of this vision lies in its concrete recommendations and the implications for a plan of action.

The TIRS vision provides us with opportunities to empower science teachers through distributed leadership. First, empowerment includes having teachers assume responsibility for achieving tasks. As teachers, we have a responsibility to move students to higher levels of achievement in science. When we are empowered, we develop new knowledge, skills, and attitudes, resulting in greater competence in our work and more impactful personal and professional outcomes. Empowerment also provides us a sense of community. For example, when leaders collaborate with the common purpose of implementing new science programs, we develop a sense of community and collegiality.

Second, achieving the new vision of science education will require leadership throughout the entire education system. I caution against placing the entire weight of realizing the new vision squarely on the shoulders of elementary, middle, and high school teachers. Instead, distributed leadership among individuals, organizations, and educational agencies is essential. We all have a role to play in contributing to the common purpose of achieving the new vision of science education for all students.

The empowerment of science teachers at all levels of K–12 education and distributed leadership across the educational community are aligned to TIRS because the focus is on implementation of new state standards for science. We now have our *purpose* from *A Framework for K–12 Science Education* (NRC 2012), our *policies* come from the new state standards, *programs* are being established, and instructional *practices* are how we as teachers contribute to the new vision. It is an exciting time because new state standards have perturbed the system. We can view this disruption not as an impediment but rather from the perspective of opportunity—new ways of thinking about science teacher leadership.

Although leadership in science education occurs in diverse contexts, it requires one to reflect on the knowledge, skills, and abilities that are necessary for one to lead. Leaders should ask themselves questions such as "What is my vision?" and "Where am I leading my students?" and "How should my students be learning science?" The questions are worth the time and effort to formulate a thoughtful response.

Leadership by science teachers involves getting colleagues, administrators, and school boards to do something they are reluctant to do—transform new standards into programs and practices. Regardless of the level of our classroom and our unique situation, we are all part of the larger science education system aimed toward improving science teaching and learning.

A leader of science education works to transform and bring improvement. With each person assuming responsibilities for leadership, the burden on others is reduced, and we have taken a constructive step toward transforming innovations into realities in science classrooms.

RODGER'S INSIGHTS AND INTERPRETATIONS

Kenneth is still in his middle school Earth science classroom on a daily basis, and as a midcareer teacher, he has achieved considerable recognition.

The TIRS vision encapsulates a vision of change for science teachers. By sharing this vision, Kenneth brings to science teachers not only an understanding of the need to change based on new state standards for science but also concrete ideas and processes to achieve the vision. He discusses the importance of distributing leadership, empowering teachers, and translating standards to programs and practices.

Kenneth reminds teachers to ask and answer essential questions about their vision and how this vision connects with their own teaching and their students' learning. He reminds the science education community that the task of reform is large and relates to all of us—who can be leaders at any level.

Leadership as an Early-Career Elementary School Teacher

CASSIE BESS

Cassie Bess is a sixth-grade teacher in the Solana Beach School District in Solana Beach, California.

NOTE FROM RODGER

I spent a day visiting Cassie Bess, which gave me the opportunity to observe her leadership in action. She began with a meeting of sixth-grade teachers at which she demonstrated collegial, subtle leadership by asking questions, presenting ideas, and providing insights about students, science, and strategies for teaching. During class time, she taught lessons covering science and math content. We joined other faculty for lunch and ended the day with a fire drill! I could not have envisioned a better way to get a peek at the leadership knowledge and abilities of an early-career classroom teacher.

Cassie teaches sixth grade (all subjects) in Solana Beach, California. She earned a bachelor's degree in child development and psychology from San Diego State University and a multiple subject teaching credential from California State University, San Marcos.

My Story

Every person has a unique story. Mine begins in a small town in upstate New York and has brought me to my current role as an online sixth-grade teacher in San Diego County during the COVID-19 pandemic. I recently received the opportunity to reflect on my childhood years and unpack how they got me to where I am today. When Rodger asked me to write a piece on leadership that looks at the roles I had from my youth to the present day and how these have influenced my current pedagogy, I immediately thought that I may not be the right person to ask, as nothing in the realm of leadership came to mind quickly. I hesitantly replied, "I'm honored, but honestly I'm not sure I have much to contribute." We talked a bit more about the opportunity, and as excited as I felt, I still couldn't scrape anything out of my memory bank that pertained to leadership, especially from childhood. *Leadership* is a pretty weighted word, after all. I started telling Rodger about some memorable programs I was a part of and sports and groups that shaped my

childhood, and he replied, "Yes, that's exactly what I am talking about." I thought he might be joking! I never thought of babysitting the neighbors and my cousins or being a Sunday School acolyte as leadership roles. But the more I thought about it, the more my understanding of the word *leadership* evolved. I am delighted to share my pathway into science education and how leadership inadvertently lit the road ahead. I will highlight the role of three guiding principles—dedication, self-reflection, and connections—as they have influenced my childhood growth, my teaching practices, and my leadership.

Early Years and Dedication

In my elementary years, I became fascinated by Olympic gymnasts and their abilities to flip around all over the place. My mom was a gymnast, and I knew gymnastics was in my blood too. I dedicated hours to the gym, dancing, and cheerleading for much of childhood. I attended camps every summer, played in my backyard with friends, and enjoyed every minute of competition. I stuck with gymnastics through part of college and have the fondest memories of the competitive atmosphere, drive to improve, and willingness to put my whole heart into something I loved so much.

Some of the most influential experiences of my life were during my senior year when I was captain of my school's varsity cheerleading team. I felt a sense of pride and loved the social aspect it brought to my life; being captain made me feel important. At the time, I focused on getting my team to work together, thrive in competitions and at games, and reach toward their individual bests for the collective good of the team. The varsity team had won multiple championship titles, specifically in the previous decade, and I so badly wanted to continue the tradition. For years, our team had shown our town that cheerleading should be a respected sport and that each member was an athlete. I wanted my legacy to strengthen this idea and strove to support my teammates in building our legacy. Our successes and failures were as a team. Looking back, not taking credit for our successes and instead being humble was no easy task for a teenager. Taking responsibility for the team's success and maintaining my dedication were the most challenging aspects of my role as captain.

My commitment to the team frequently guides my drive to create a classroom community with each new group that I teach. Seeing kids bond after thinking they had nothing in common and creating innovative learning opportunities is much like getting a team to the Olympics, from my perspective. Dedication and team spirit lead the way!

A Self-Reflection

As I reflect on my captain days, I am prompted to think about my current students' preparation for their student-led conferences. As sixth graders, they are challenged to consider their strengths and weaknesses as the trimester comes to a close. Each student presents to their family and me about their areas of growth and topics for refinement in connection to work samples, class discussions, projects, and personal self-development. This reminds me vividly of the self-love and need for growth that I consistently needed to exhibit as a captain, especially in moments when I was ready—and definitely in times when I was not. Not only does reflecting on areas of strength and need put me back into the student role, but it also allows me to connect to my students in a whole new way. Student-led conferences are an impactful way for students to take responsibility for their own learning while having an open-door environment to grow from the exact place they stand in that moment. I am fortunate to be able to foster this opportunity with my students and hope that they use self-reflection as a stepping-stone for future growth.

Creating Connections

Creating connections with students and the team of teachers who inspire me daily is one of my strengths and something that I strive to accomplish every single day that I teach. There are many areas in which I can improve, and I am fortunate to have a large team of incredible professionals with diverse skills. My mom knows how to connect with every coworker, client, family member, friend, or stranger in the grocery store in a way that makes that person feel like their relationship was decades old and built of memories. To this day, our connection gets me through all the stress and challenging opportunities that life throws our way. Our relationship is also a staple in every celebration, milestone, and accomplishment in life. Having our connection as part of my backbone constantly reminds me about how crucial it is to form bonds with students. So much learning comes from listening and building relationships. Understanding a whole student and helping every individual understand the real-life meaning in educational lessons is the true heart of teaching. Connections start from the first day of a new school year and often continue well into a student's education and future. Learning about and including students' families, building projects and learning opportunities with choice based on their interest, and asking for their feedback and what they care about are just the beginning. I truly believe connections are the most important building blocks and necessary for successful leadership.

Putting the Principles Together

During the first summer of the COVID-19 pandemic, my district offered students an opportunity to continue learning through meaningful projects. A colleague and I created a virtual learning choice board for sixth-grade students to choose activities that they were passionate about, all based on the theme of plastics in the ocean. During the first weeks, they learned about the striking statistics behind the current state of our oceans with regard to plastic pollution and the effects on wildlife, took part in a beach cleanup, made art out of "trash," and more activities. Their learning concluded with an opportunity to create a public service announcement or write a persuasive essay. In efforts to reach a larger audience with the students' culminating projects, we added a little twist. Local professionals with a similar passion were asked to judge the student submissions, and we gathered prizes from a nearby surf shop and jewelry makers who use ocean plastic. The ideas all stemmed from recent students who, while learning about human impacts before the pandemic hit, shared stories about why they and their families enjoyed the unit and the plans they had to make a positive difference concerning climate change (many specifically connected to plastic pollution in the ocean). By adding a real-world audience and a contest, students were engaged over summer and worked to make a positive difference, even during a global pandemic! This is just one example of how connecting to students' interests can help learning thrive and become meaningful and exciting. In this project, students developed dedication, self-reflection, and connections.

Conclusion

I hope the word *leadership* has a brand-new meaning for you and that you are inspired to use your dedication to the profession, reflect on your own practices, and connect to teaching in a whole new way. My observation on leadership is intended to be a brief share about the beginning of my career and what drives my passion. I believe we have a job that can be full of creativity and uniqueness if we allow our hearts to explore it!

RODGER'S INSIGHTS AND INTERPRETATIONS

In my observations of her teaching, Cassie was clear about her goals for the lessons and showed flexibility in the means of achieving those goals as a result of her recognition and acceptance of individual students' needs.

Cassie was very active as she taught her class of 22 students. She seemed very comfortable directing the class as they began work for the day. As she set up lessons for small groups and answered questions, she demonstrated leadership, a command of both science and math content, and an understanding of the purpose and direction of students' work. As she worked with students on science projects, she led with coaching questions, suggestions for resources, and next steps as students encountered challenges with applying technologies and analyzing data. Cassie worked comfortably with both individuals and groups, being clear about her educational goals and efficient and direct in her interactions with students.

Different groups culminated their day by giving brief reports on careers in science and biology, such as microbiology, epidemiology, and geology, covering what the individuals did, an example of questions they asked, and how they went about investigating the natural phenomena and gathering data to answer questions.

Near the end of the school day there was a fire drill. All students had to leave the classroom in an orderly manner and go to an assigned location in the schoolyard. As the students gathered, two boys began pushing and shoving each other until one boy hit the other with a direct blow to the nose, which resulted in a significant amount of blood.

Cassie immediately separated the boys and stopped the bleeding. She asked each boy for an explanation of what happened. Later, she had to complete a report for the principal and call both students' parents. Her leadership in this situation was direct and immediate.

With her colleagues, Cassie showed subtle leadership by advocating for standards-based approaches, collaborating with her sixth-grade team, and demonstrating different uses of technologies and instructional materials.

Cassie has participated in professional development with her school and district to improve her curriculum and instruction. She has advocated for following the *Next Generation Science Standards* and tried to align her school's program to those standards by preparing and sharing a model unit of instruction with colleagues. These are examples of leadership that, it seems to me, are appropriate for an early-career teacher.

Cassie's priority of recognizing students' needs is aligned with what has been identified as person-centered leadership, as she communicates her goals and shows flexibility in her plans based on students' responses.

I later asked her to reflect on her leadership in the context of the day and her goals. Here is her response:

> *Across the years, I had not thought about how various experiences had impacted me and contributed to my leadership as a classroom teacher. My first priority is to recognize students as the most essential piece of the teaching puzzle. What is best for the kids, and what are their needs?*
>
> *A second issue is communicating my goals and designing experiences for students that have an end in mind but have flexibility based on students' responses. It is something like the Chutes and Ladders game. You have a finish line in sight, but sometimes you have to slide back down and reteach before you can continue.*

4 Achieving Effectiveness as a Teacher Leader

CYNTHIA ROUNDS

Cynthia Rounds is a middle school science teacher in Fullerton, California.

NOTE FROM RODGER

During a workshop for science teachers in the Fullerton (California) School District, I noticed one teacher with enthusiasm and motivation to learn science and new teaching strategies. After asking Dr. Robert Pletka, superintendent of the Fullerton School District, about the teacher, I learned she was an outstanding science teacher and a leader.

Cynthia Rounds teaches seventh- and eighth-grade science and has been involved in numerous science initiatives, all of which have contributed to her development as a leader. She earned a master's degree in education from California State University, Fullerton.

The Early Years

We all gain confidence in different ways—I gained mine at home. My father was a math teacher and my mother managed the home and four children. We were always expected to pitch in with maintenance and repairs when needed. On Saturdays, we were assigned a task (such as doing yard work, painting the house, or even repairing the car with Dad) to complete (including any problem solving) and wait for inspection; we were expected to do it right! A good work ethic was ingrained in me. We also went camping for four weeks every summer, and I packed my own clothes and toys from an early age. Leading family hikes, climbing, and exploring provided learning experiences and fostered independence. My self-esteem developed as these experiences helped me think, self-evaluate, and problem solve.

My youngest sister was born when I was 10. I shared a room with her and I learned all about caring for babies. I became an in-demand babysitter around town, which made me feel responsible and independent. These experience increased my confidence and abilities that I would later use in the child-teacher relationship.

My family was involved in church. When I was in junior high school, my dad asked me to help him teach Sunday School, and I eventually taught my own class every Sunday. Although I was reserved and introverted and did not want attention or people looking at me, I didn't mind small children. I learned to speak up when needed, which made me feel heard. When I was in high school, my youth group leaders saw something I did not see in myself. Mr. Clark, my youth pastor, told me I would make a good leader. Mr. Clark and other leaders invested in me through training, building, and challenging me, and I soon led a small group of girls. Each spring break, the high school and college students would go to Mexicali, Mexico, to put on programs for the kids. Each team member was responsible for one or more activities each day. This responsibility pushed me to lead a group of children in front of my peers in a safe environment. I went to Mexico six times, and these trips showed me that working with children was fulfilling and I could organize activities and lead my peers.

After graduating from high school, I applied for a summer job at a large camp in Northern California. I was the head baker for 900 campers, which was different, fun, and a major responsibility. I went back again the next summer and returned home between summers to attend the local community college, where I began to take classes for a teacher credential. My best friend got me a job in day care, where I gained experience with budgeting, creating activities, communicating with parents, and handling discipline.

I earned a bachelor of arts degree and applied for the teacher credential program at California State University, Fullerton. Finally, student teaching began. My first semester of student teaching was in a sixth-grade classroom with a young and energetic master teacher. I learned a lot from her. The work was hard and rewarding at the same time.

I was placed in a third-grade classroom for my second semester of student teaching. The teacher was seasoned and older with many great ideas, but she was controlling and moody. She corrected me constantly on whatever I did while either of us taught. At the end of my first week, she told me that I was too soft spoken for this job and I would never make a good teacher. I really wondered if I had made a mistake. Then I became angry and felt powerless to do anything about it. I wanted out of that room and asked my supervisor if I could change to another master teacher. My request was denied. I decided to smile, suck it up, and get it done. It was a long three months, and many tears were shed. I was determined to not let this happen to another student teacher. After I turned in my evaluation, I met with the principal and my professor separately, explaining my perspective and requesting that they reconsider using this master teacher in future years. I was so nervous to deliver that message, but it was well received and I left satisfied that I had done the right thing. Seven years later, I was teaching at that school again with that teacher still on staff, but this time I was not in a powerless situation. She knew that, and we both acted professionally. I was glad I took the high road the first time, as a great life lesson for me was to always leave in good standing and act professional.

Experiences in Science Education

I began my first teaching position at Commonwealth Elementary in the Fullerton School District. I knew by then that I liked to teach older students, and I felt sixth grade was a good fit. I was so enthusiastic and jumped into the classroom with both feet. The first year was exciting and difficult. I felt comfortable teaching the curriculum, with the exception of science, which I loved until high school and avoided it thereafter. The science textbook in my classroom was old and students were not engaged. After buying books and researching, I slowly started to add some fun experiments. I became the site science representative for my adjunct duty with the hope that this role would connect me with other teachers in the district who could support me.

I was one of two new teachers that year, and we stuck together. The other new teacher loved science and helped me with ideas, such as Family Science Night. We got a book, planned the event, and put about 15 experiments in the multipurpose room for that night. Families came and did the experiments together. Family Science Night was a success and soon became a yearly tradition. I realized that science was quite fun and the students enjoyed doing it. It was becoming clear to me that giving students experiences in hands-on, inquiry-based science motivated them to learn.

I still needed some units to complete my multiple subject teaching credential, so I was on the lookout for ways to juggle a new job while taking classes. I saw a flyer from California State University, San Bernardino's (CSUSB) Geology Department in the lounge at school. They were looking for teachers and offering hands-on summer science training. Even better, earning credits was an option. The geology program was amazing. We went whale watching, spent a night in a lighthouse, saw the rings of Jupiter, and drove all over California looking at landforms and faults and gathering rock samples. I walked away with amazing science lessons, samples, and pictures, as well as a grant to implement these lessons in my classroom. I was asked to return the next year as a paid co-teacher. That next year was bigger and better, as I became comfortable teaching my peers in this informal, safe, and fun environment. I had energy and enthusiasm when teaching topics that I experienced firsthand, and administrators began to notice I taught science in a fun way. I became confident in teaching science and started to share my ideas within the district. I was asked to co-present CSUSB's grant program at the California Science Teachers Association conference, my first of many experiences presenting at conferences.

Implementing a New Science Curriculum

A few years into my career, Arnold O. Beckman decided to invest in science education. His company, Beckman Instruments, revolutionized the study and understanding of human biology. The Arnold and Mabel Beckman Foundation developed the Beckman@Science Program, a K–6 science education initiative to serve 300,000 elementary school children in Orange County, California. By 2003, the foundation had spent approximately $14.5 million on a science education program that has captured children's natural curiosity and stimulated their interest in science through hands-on, inquiry-based science learning experiences.

My district wanted to join this amazing opportunity. A small committee would serve as an advisory board for the program and would consist of an administrator, three teachers, and a local science professional. I was one of the few in the district who was involved with science and was asked to join the committee. The Beckman@Science Program was well planned and executed. A team from each district in the county participated in a week of intense training. We were trained during the day in the Beckman@Science meetings and then sent off as a district committee to work on a proposal. We left with a new vision and a grant outline, which became a grant proposal that we presented to the Fullerton School District Board of Education, then to the Beckman@Science Program. Our district was selected, and the adventure began.

Our team met weekly at the district level and monthly with the county teams. The grant money was for a three-year period and included curriculum adoption with kits, a plan to refurbish consumables, and intense training for administrators and teachers. We chose our hands-on, inquiry-based curriculum for all kindergarten through sixth grades. A team of lead teachers, including myself and administrators from each school, was chosen and trained on the pedagogy and use of the science kits. A building was designated for storing and refurbishing the kits, with an office for the teacher on special assignment to manage daily implementation of the plan. Every teacher in every grade was trained in a hands-on, inquiry-based science curriculum. I found myself training principals and teachers and working with community professionals and volunteers. The original grant was for three years, but the foundation funded a few more years as long as goals were met and the district slowly shifted toward self-sustainability.

Beckman@Science also began a program that trained teachers with the intent to put science leaders in each district. This was a countywide program that consisted of intense training in the summer and monthly meetings throughout the school year. This grant program continued in our district for many years because of the long-term plan. We kept training teachers, filling kits, working with the community stakeholders, and teaching kids science in an effective way. This program changed the way I taught and thought about science education. I was morphing into a science teaching leader.

Somewhere in all that excitement in my career, I had my first baby. During that first year, I could not do everything I had been doing before she was born; I was looking for a change. A position opened at a neighboring school in the district when a teacher retired. This school had a special science teacher who taught science to every student. The school had a science room, and classes would come in one at a time. This was also a 60% position—perfect for me. Since I had been implementing the Beckman@ Science grant, I had experience in all grade levels and was familiar with all the kits that were recently adopted. I also had a few days a week at home with my baby. As with any new position, I worked more than the part-time hours. District administrators began to film my teaching to help train teachers in the newly adopted curriculum. I invited teachers on-site to join me with their class to learn how to teach the lessons, so that they could teach the lessons they observed the following years, when I would do different lessons.

Near the end of my first year in this new position, the director of personnel informed me that I was not qualified for this job and needed a science credential. I had to begin pursuing a credential or find a new position. The end of the school year was nearing, and I did not know what I would do. I did not want to go back to school. I was working part-time and could not really afford to pay for school unless I went back to a full-time position, and the whole point of accepting this new position was to spend more time with my family. A few weeks later, I found a flyer in my mailbox from California State University, Fullerton. The science department was offering teachers science classes on Saturdays in an intensive three-year program to pursue a science credential. The best part was that the program was funded by a grant and would not have any expenses for participants. I applied and another journey began. A cohort of science-loving teachers participated for the next three years. At the end of the program, I defended my thesis a few weeks after the birth of my son.

The district was about to finish building a new school and announced that it would no longer be an elementary school but instead would serve grades K–8 to alleviate overcrowding in the junior highs. The district also said this new school would be a state-of-the-art science and technology school. The Robert C. Fisler School was still being built, so I was consulted about layout, closet space, and storage and tasked with implementing technology. I spent my summer researching, asking peers about their needs, and making order lists. My list was the ultimate wish list. A week after submission, I met with the principal, who said the district board would not approve such a list because it was not what other schools had. I asked her respectfully if the board wanted a 20-year-old lab or a state-of-the-art science and technology lab. How can we have such a school with little resources and lack of technology? I believe my involvement and leadership at the district level for all those years put me in a position of respect so that my request was considered seriously by all involved. I got word following the next board meeting that every item on my list was approved and would be purchased.

Integrating Technology in the Science Program

The school district formed a partnership with Apple prior to opening the new school. Teachers began training over the summer and throughout the first years on the Apple programs. None of us knew what a 1:1 laptop classroom looked like or how it worked. Teachers spent many hours discussing and sharing best practices. The district (in response to my request) had purchased USB microscopes and probes to use with the laptops. I began to write labs and lessons for a paperless classroom. I was the only science teacher at my site, so expectations were high.

The following year got better. Students were catching on to the technology, and teachers were integrating it into teaching. Science was fun when we used technology. People from all over the world wanted to see what a science and technology school looked like. I became more comfortable with adults popping into my room. I was getting attention and must have been doing some things right because Apple invited me to apply for admission to an elite group of teachers. I became an Apple Distinguished Educator, one of approximately 100 others (at that time) from around the world. I did not think I was good enough to be in the presence of such great innovative teachers; I found out that almost everyone there felt that way.

The fun part about using technology and science together is that both are always advancing and changing. I never taught the same concept with the same tools the same way after the school's opening year. My new motto was "What if? Consider the possibilities." This motto took me to so many amazing places in teaching science. I was constantly trying new things, but this often took money, so I began to write more grants. I was on the crest of a technology wave, and many organizations showed interest in my ideas. My brother (a great businessman) told me to use my position and recognition to get what I wanted. Sometimes I would get to use things for free just by asking, because we were a top-tier school and affiliated with Apple. Most of the time, however, I would put the time in to write a grant. Grants led me to many firsts. For instance, I was the first in the district to get an interactive whiteboard. I loved it. I created science lessons to use on it and soon presented at conferences for the company. Fast forward a few years, and there was a board in every classroom on campus and most classrooms in the district. The same thing happened with a learning management platform.

When trying new ideas, technology, and science equipment, failure is inevitable. Prior to my middle school position, I was the teacher who had to know more than any of my students. I thought a failure in the classroom would make me look weak and like I did not know what I was doing. I had it all wrong. Failing is part of growing and learning. If you are not willing to take a risk, you will not progress. When something flopped in the classroom, I would apologize and own it. We would discuss it and try to come up with a solution or alternative together. It was okay that some students knew more about technology than I did. We learned together and from each other. Many times a year, I would try new labs, technology tools, and applications. I always introduced them to my classes as something new that I needed them to try to use and help me improve. We would go through the journey together. Sometimes I declared it a total flop, and we would start over with a new way.

Paving a new path naturally brings attention. Administrators and teachers began to pay attention to what I was doing and would inquire about it. A publisher asked me to write a science book, and I was invited to present at meetings and conferences. I was also one of the first educators in the United States to use an application from Agents of Discovery, a start-up company, in which questions or tasks pop up on a device when you enter a certain GPS location (like Pokémon GO for education). When I introduced the app to my students, we played a mission as a unit introduction, then debriefed with the company's chief executive officer in a videoconference. During the unit, the students learned how to write and make their own mission and how to publish on the platform. This process involved bringing our success and failures to conference calls with the programmers so they could add to and change the application to make it work better for students.

Forming a Foundation

A few years after the opening of Robert C. Fisler School, it became clear that it was expensive for the district to maintain a ratio of one computer for every student. The question of equity between schools was raised. The district supported the program but pushed for more funding to come from the school. We had something great and wanted to keep it. We also realized that we needed financial assistance for equipment, apps, and technology support. A not-for-profit foundation called the Fisler Foundation for Science and Technology was formed for the school. It made sense for me to be on the founding board since I was the science teacher on-site. Serving on the board allowed me to give direction and guidance in what to fund for science schoolwide and taught me a lot about nonprofits and raising money. Once again, I partnered with parents and professionals in the community, and the foundation has raised a lot of money for our school. I had started an annual Family Science Night that included a hands-on activity, a short movie, and a teacher and volunteer in each classroom. Over the years, Family Science Night has grown into Science Week, with assemblies, activities, and a science night. After the foundation was formed, it sponsored Science Week. The foundation also funds Science Olympiad, and each year it provides grants to teachers who make dreams come true.

I still teach science at this school. Over the years, there have been many tours and visitors and lots of progress, and the students continue to do amazing things. We continually try new science tools and applications. We code and we do a bit of gaming. We apply science to real-world problems. We are curious and inquisitive about our world. My teaching is always evolving.

Looking back, I can see how I was blessed with many opportunities during my career. Each opportunity were more complex and challenging than the previous one. I probably would not have taken each opportunity if I did not have the experiences of the one before. I was scared to pursue many of those paths but believed I could do it because I had learned and succeeded with the prior tasks and was supported and encouraged by those around me. Coming out of each experience increased my skills and confidence. I did not see myself as a science leader initially; becoming a leader was not my goal. I just wanted to be the best at whatever I did and find joy in the journey.

Advice for a Young Teacher Leader

1. A good work ethic will take you to your destination. A leader is passionate and passion initiates action and attracts others. Endurance is just as important. Be the person who gets it done. Problem solve and find new paths when hitting a roadblock. Always move forward by learning and reinventing yourself.

2. Be bold and integrate new things. Surround yourself with and seek out other "first adopters." Consider the possibilities. Keep it fresh, and build and morph your labs and lessons.

3. A teacher leader is prepared. Take time to know the foundation on which you are building science programs. Know your state standards and science teaching's best practices. Write meaningful, fun, and concise labs. Be organized in the presentation and execution of lessons and labs. A prepared person is naturally more confident. Stop and look at the whole picture of where you are now and where you want to go. Take time to know what you believe is right for students and for you, and let that influence your decisions. Always find the *why* in what you do. When you know that, you know the direction you want to take and what you need to get there. Have facts, statistics, and wish lists ready. Be ready to seize new opportunities.

4. Seek those who will share with you and support you. Make connections. Find people you admire who make things work in their classroom. Learn from them. Surround yourself with positive people, because you will fail and get discouraged, and others may tell you that you cannot do it. It is not the failure that is important—it is what you do after that failure. Learn from it and keep going.

5. Networking is important. Get to know administrators, parents, and community members. Join committees, chats, and professional science associations. Attend site and district meetings. Never burn a bridge. Be a cheerleader for others and invite them to get involved. While you are networking, share; we all have something to contribute. You are a great teacher—believe it. Find your strengths and share them. All great teacher leaders think they are not as good as others they see in the room or on social media. If you don't believe in yourself, others will not believe in you. Teaching is not a solo profession. Get out there with enthusiasm. Remember to balance your confidence with humility.

6. In all that you do, know your audience. Whether you are speaking to children or adults or writing a grant, make sure your words are pertinent. Know students' strengths and weaknesses so you can meet their needs. When writing a grant or speaking to parents and adults, share how they can help you reach a goal.

Conclusion

I have always been enthusiastic about my science teaching. Enthusiasm is vital for effective teaching and leadership. In the beginning, I didn't think of myself as a science leader; I just wanted to become the best version of me. I am a reflective person and an introvert. Early on, I realized that I needed to be heard and speak my convictions. If I did not, I would be hurt or have to live with others' decisions. I would much rather live with my decisions and my vision. A science leader does not follow a path; she chooses her path. If the path is not there, then pave your own and have fun.

RODGER'S INSIGHTS AND INTERPRETATIONS

Cynthia's professional journey is like an odyssey, filled with notable experiences that contributed to her knowledge and skills as a leader. For instance, she adapted to accommodate her responsibilities as a parent. She also integrated new technologies into her classroom and eventually the district. Cynthia developed from a person who avoided science in high school and college to one who is a credentialed science specialist in her district.

She offers sound advice for young leaders: Develop a work ethic to achieve your goals; have the courage to integrate new strategies, technologies, and programs; networking is important; be prepared for your leadership responsibilities; seek mentors as models from whom you will learn and get support; know your audience and interact with them as individuals of integrity and worth. Most important, she notes that you should continually keep in mind the big picture of where you are and where you want to go.

5

From the Classroom to Leadership and Back

FELICIA RYDER

Felicia Ryder is a science teacher at Wangenheim Middle School in San Diego, California.

NOTE FROM RODGER

After graduating from San Diego State University (SDSU) with a degree in biology and health sciences, Felicia Ryder began teaching at Horace Mann Middle School in San Diego, California. In her initial years of teaching, she extended her experiences beyond the classroom as an aerobics club teacher, an academic portfolio assessor for the district, and as part of the K–12 Alliance Science Professional Development program. In 1996, Felicia completed a master's degree at SDSU.

Early Leadership

One of my earliest recollections of being a leader is from age 10. My best friend, Karin, and I decided to organize a neighborhood variety show (kind of like Donny and Marie or Sonny and Cher). Karin verbalized lots of ideas while I had the vision and blended our ideas. As I think about it now, this memory makes me feel happy, and if I analyze the experience, my happiness and fulfillment seem to arise from the way in which I led us to accomplish our vision. I was involved from the ground up as a decision maker, creator, and actor. I handled public relations, snacks, and cleanup.

The most memorable part? Being onstage as Kiki Dee to Karin's Elton John, belting out "Don't Go Breakin' My Heart" to the neighborhood kids sprawled on the driveway. I believe this is where my leadership style emerged. I don't like telling others what to do; I like doing the things right alongside those who are learning.

As far as early leadership experiences, I didn't have any of the typical ones. My parents worked hard to raise my sister and me and get us to school. They provided us with a wonderful childhood full of memories, but extracurricular activities were not part of this time.

Fast forward to high school. My friends were my world, as they had been since middle school. Lucky for me, my friends were a really good influence. I applied for Associated Student Body (ASB) on the

24   NATIONAL SCIENCE TEACHING ASSOCIATION recommendation of good friends and got in! I was amazed because this would be my first official leadership experience. (I tried out for cheerleading twice and didn't make it—ouch!) My leadership style grew in ASB, though I still wanted to create and do, like building a float for weeks after school and riding it in the parade, all while socializing with my friends.

In high school, I took a formal leadership class and recall themes of teamwork and trust. I remember a self-analysis project in which we answered these questions: Who are you? How do you connect with others? Where do you see yourself now and in the future? I grew in that class. I enjoyed and excelled at collaborating with others. I listened intently when my classmates shared personal stories of triumph or heartache. I practiced presenting to my peers and got pretty good at it. In fact, leadership institutes I've participated in as an educator have included much of what we did in that first leadership class.

Development as a Professional Teacher

Finding my way to a career was not straightforward. I enrolled at a junior college in San Diego and signed up for general education classes that I needed to transfer to SDSU. For what degree, in what area, I had no clue. I worked part-time and earned good grades, then needed more money, so I worked full-time. I thought I liked accounting and might go that route, so I took a higher-level class, which I loathed. I finally landed in a class that sparked my interest: nutrition. This class was interesting! The human body does *what* with vitamin D? "This is it," I thought. I wanted to be a nutritionist or something similar that involved people and diets.

SDSU accepted me into the College of Health and Human Services. I devoured classes in anatomy, physiology, and biology and developed a passion for learning and collaborating with others. Soon I was told I needed to pick a career and a specific goal, and after meetings with university advisers, it looked like public health was the arena. At the time, I couldn't visualize this as my career, so I kept my mind open and continued in these classes, one of which required volunteering in the community. I was referred to a list that included a high school program for teen girls who were pregnant (Expectant Teen Classroom). The girls needed tutoring in biology, so I signed up for that job.

My volunteer hours flew by as I showed the soon-to-be moms how to complete Punnet squares, explained what the mitochondrion does for the cell, and helped them understand what the letters *A*, *T*, *G*, and *C* have to do with DNA, among other topics. I supported them with writing and studying for tests. I didn't think about how I was teaching them; I simply tackled the material. I didn't have any methods. I recall drifting eyes and expressions of boredom, and I always asked them to tell me how they would say or do something. Then we'd add the scientific terminology and process it. The more they could talk, the less annoyed they seemed, because they just wanted to be done with school.

But both before and after any of this took place, I listened to them when they told me their life stories because I wanted to connect with them. Their backgrounds were so completely different from mine, but I felt a strong need to be relatable in even the tiniest way, so I listened with eyes and ears. I also turned red when they thanked me for helping them earn an A or B on an assignment or a test. I'd well up when one left school to have her baby.

My Decision to Teach

Teaching in the ETC program was a life-changing experience for me and led me to science teaching. Each day that I left that classroom, I was fulfilled. Sharing my knowledge with these young women was uplifting and humanizing. I knew then and there that I wanted to teach something that helps humanity, like nutrition or disease prevention.

I finished my major in biology and earned a secondary teaching credential. I completed my student teaching at a high school and a middle school and learned that teaching middle school students is fun! Middle school kids are hilarious. I secured my first full-time teaching position at an inner-city middle school. My heart was full.

Emerging as a Teacher Leader

Leadership came into the picture slowly, in small pieces over time, and grew with each new connection. Being in education, you get asked to do lots of things for the greater good—meaning for no pay, after hours, and possibly with some free materials, if you're lucky. But the bonus was being able to network and gain allies and mentors. As a young, new teacher who was single with no kids, I took on a lot of those tasks. I developed into a person who liked to be on the cutting edge of teaching, who wanted to know the newest methods, who was willing to teach the unit and gather the data to send back in exchange for free books or curricular materials or even just the experience. That was probably one of my first educational leadership experiences—taking on projects with other educators and curriculum designers to help gather data or test or pilot a curriculum.

Another leadership opportunity came in the form of presenting to my peers. My administrator asked me to share my classroom management techniques with staff. I remember fretting about that presentation. I really wanted to share my ideas, but the thought of standing in front of my colleagues was mortifying. I can still picture the people who actually paid attention and were involved—and the ones who sat in the back and talked. I'm sure my presentation had some good points, but I'm equally sure that I presented to my peers differently than I did to my students. I didn't know it then, but that was a mistake.

Although I had no formal pedagogy training and knew nothing of the 5 Es, inquiry, or phenomenon-based teaching, I did know that my students needed something to spark their interest, to do, and to read and reinforce the information, as well as a way to wrap it all up for themselves at the end. Why I didn't structure my first professional development session to my peers that way, I will never know, but it was a great learning experience. All it took were a few nonparticipants to make me realize that if I did the session as I did for my students, my peers would have understood it.

The yearning for "them" to understand has been with me ever since. It's just that the "them" has changed from students to peers at my school to colleagues in my district to professionals across the state and country and in different venues.

What Became Important and What I Learned as a Leader

As my teaching career progressed, I became involved in two organizations that influenced my educational leadership, the BSCS (Biological Sciences Curriculum Study) National Academy for Curriculum Leadership (NACL) and the WestEd K–12 Alliance. NACL was a two-summer program in which I learned not only how to improve my own pedagogy through inquiry teaching but also how to systematically compare curricula using a process called Analyzing Instructional Materials (AIM). In the AIM process, a team of educators collect data that show how likely a curriculum is to enable students to connect ideas that lead to higher levels of learning. We engaged in the process authentically, bringing back evidence and documentation we could use to make data-based decisions about curriculum adoption. We also brought back knowledge of the process and the goal of training others in our district to understand and participate in the AIM process.

In this phase of my leadership journey, I was not the first to speak; I listened. I would not take risks but participated fully in tasks with clear procedures. I wanted to learn but was uncomfortable in the leader role. Being surrounded by colleagues who felt the same as me was such a relief. We went through it together, and some of us got really good quickly while others moved quite slowly. I was somewhere in the middle, trying to figure out my leadership aptitude.

Part of being in the NACL group included a commitment to work with teachers in our schools and district to build their capacity to make good curriculum selections. I focused on the details of what I would learn and the picture of that knowledge as applied across our district of 130,000 students and 1,500 middle and high school science teachers. Elementary teachers numbered in the thousands. In my mind, that picture looked amazing and motivated me to think about what it would be like if all the science teachers had an exceptional curriculum and understood how to implement it. I also considered what it would be like if I could be the person to help those teachers. That second thought was a strong hook. I'd been given the privilege to effect change in a much greater way than in just my own classroom. That eased my stress enormously and enabled me to say, "So what if it's scary presenting to adults? Think about the results! Think about how many more students can be affected if I can reach and support even a few educators." These early thoughts were really what got me over the anxiety that overwhelmed me, as it does so many potential leaders.

What is the vision? Is the vision for the organization defined? If not, it needs to be, and if so, is it one that you're on board with? Is it a vision that you can see yourself helping to actualize, and will doing so fulfill part (or all) of what brings you joy about your career? If the latter part is true, my recommendation is to get on that boat and begin rowing. Take that step. Don't overthink it—there will be plenty of need for details later. For now, keep your eyes on the beautiful vision and go for it! The education community needs you.

We created professional development based on the concepts of the sessions in NACL. I committed to spreading knowledge and skills and took on a leadership role. I was still somewhat anxious, but the information and skills we had learned and were ready to share made this responsibility exciting. I was grateful that some had seen the potential in me and put my name out there for this role. I was also thankful and relieved that the professional development program we worked on at home was collaboratively created and co-facilitated.

The K–12 Alliance also honed my skills as an educator and leader. In the K–12 Alliance, I learned as a student, teacher, and facilitator. With structured guidance, I helped create professional development scripts and materials for other educators. At some point during my involvement with these two organizations, I learned about the BSCS 5E Instructional Model and the concept of inquiry-based teaching and learning. My career was forever changed, and once again I saw how this knowledge could inform, affect, and improve science education as a whole.

I became a middle school science resource teacher for my district after 10 years as a classroom teacher. I worked with a wonderful group of science educators and collaborated to create and present professional development on a variety of topics: inquiry-based lessons, 5E instructional design, lesson study, curriculum implementation, pedagogy for beginners and veterans, and more. I found that my style of leading was straightforward and no nonsense, but at the same time I left no person behind. I brought with me that vision of an informed future improving students' lives, which helped quell the lurking butterflies. I also brought the lessons I learned as a teacher and a teacher leader in NACL and the K–12 Alliance.

What became important to me as a leader were larger, umbrella ideas as well as some smaller points. Here are some strategies I used in every professional development meeting:

1. Post the agenda and stick to it.

2. Create norms as a group, and have somebody monitor those norms.

3. Honor the expertise in the room. I can learn from you as much as you can from me, so let's do this together.

4. Use the "student hat" and "teacher hat," meaning participants will at times learn as a student in their class does and at times learn as the teacher.

5. Follow the 5E Model, which is based on the research on how people learn—all people, not just kids.

6. Allow participants time to connect and reflect.

7. Use backward design planning.

8. Model the pedagogy in every session.

In all of our professional development sessions, we would collect feedback at the end in a simple way: a T-chart on a poster, with one side labeled GOTS and the other labeled NEEDS. Each participant would place sticky notes on each side with their ideas, what they got out of the session or learned, and what they still needed. This approach was powerful. It created feelings of gratification and achievement and provided us with concrete ideas for our next steps.

Lesson study became a districtwide focus. I discovered through facilitating lesson study that I just could not keep myself out of the creative planning process. It was difficult to solely guide and not also create. In fact, my favorite lesson studies were the ones in which I was co-facilitating, co-creating, co-teaching, and then co-analyzing the student work with participants. As I look back at the many professional development sessions that I presented, my favorites were the ones in which I was deeply involved with participants.

The takeaway for me in this part of my leadership journey is that I recognized my leadership style, which includes a need for participants to acknowledge me as a teacher and a lifelong learner. With that established, I was at ease. I made sure from then on to share my story with participants, as well as my eagerness to work with them and my excitement to learn from them.

Through the years, I have found that the most important requirements for leadership are collaboration, communication, connection, and purpose. Participants must understand the purpose of the professional development. As I led professional development and shared what I did and a bit about my background and values, I would see shoulders relax. When I gave participants the chance to share their own stories, they began to make more eye contact, to connect with each other and with me.

I learned the value of a collaborative approach to planning and leading and, along with that, how crucial it is to credit those who contribute. By honoring the creator of an activity or strategy, I expanded the pool of resources for all. I gave those who supported me along the way a stepping-stone to move into a leadership role in the future, should they choose, by publicizing their work. I learned to honor the expert in the room by asking questions and deferring to those who have answers, suggestions, and ideas. It's about the learning.

Becoming a leader in the science education community enhanced my teaching. I was privy to resources that were brand new and those that weren't necessarily new but were data driven and new to me. I met other educators and scientists who I could reach out to with questions and for support, and I was offered more opportunities to continue leading and learning.

I encourage new teachers to get involved in the science teaching activities in their school and district. Join a curriculum study group, attend professional development sessions, or present something to your department. Do not try to do it all on your own—collaboration is key.

Return to the Classroom

After a number of years, I missed students in the classroom. I found I wasn't as fulfilled when not teaching them. I also found that there were some adult learners who didn't want to learn, which was disheartening. These struggles became exhausting. They didn't keep me from leading, though, and what got me through was the collaborative approach to presenting and the research-based techniques that we used.

I returned to the classroom and have stayed connected with the *Next Generation Science Standards* community in my district and nationally. I've volunteered to do lesson studies with other teachers through our county's Office of Education. I use the resources of the biotechnology community and sign up to hear guest speakers and attend field trips. I engage in all of these activities as a direct result of my time and experience as a leader.

Ultimately, the classroom is the place for me. I am a middle school science teacher at heart and am happy when I go to work every day. I still enjoy finding and sharing resources with my department and other teachers in my district and modeling how to honor others' creations and contributions. When I do so, I've observed that at the very next meeting, the person I honored speaks up a little more, leads a little better, and is a little prouder.

RODGER'S INSIGHTS AND INTERPRETATIONS

Although she is an introvert, Felicia's profile makes it clear that she had the interest and energy to overcome any reluctance in pursuing leadership. With a unique constellation of experiences, such as tutoring expectant teens, she emerged as a science teacher and leader in her school and district. Felicia's enthusiasm was complemented by her participation in NACL and the K–12 Alliance, both of which provided professional learning experiences.

Throughout her profile, Felicia presents small gems about leadership, such as that those in education get asked to do many things; teachers should become involved in organizations that form their educational leadership; leaders should always keep their eye on the vision; and it's important to take that first shot at presenting.

Felicia boils down her experiences and knowledge to the most important requirements for leadership: collaboration, communication, connection, and purpose. We would all do well to remember these when we take on leadership roles and responsibilities.

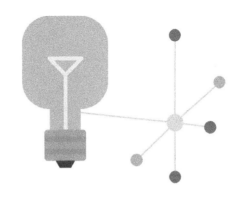

PART II

Some Leaders Leave the Classroom to Support Science Teachers

From a Science Teacher to Educational Leadership

Recognizing Common and Uncommon Opportunities

JAMES BLAKE

James Blake is the director of strategic partnerships and focus programs for the Science Focus Program and the principal of the Arts and Humanities Focus Program for Lincoln Public Schools in Lincoln, Nebraska.

NOTE FROM RODGER

In 2016, I was invited to give a presentation at the University of Nebraska–Lincoln. James Blake, the new state science coordinator for Nebraska, attended the presentation, and I saw his interest and enthusiasm. He was, by definition, a leader. In our short initial discussion, I learned that James had first been a science teacher. Since this first meeting, James has continually demonstrated the qualities of a leader and assumed positions of leadership. In October 2019, I spent time with James, including shadowing him at work for a day.

James earned a bachelor's degree from Buena Vista University in Storm Lake, Iowa. He earned a master of science degree and a master of secondary teaching degree, as well as a doctorate in educational administration, all from the University of Nebraska in Lincoln, Nebraska.

My Story

I offer my story to help identify some common markers along the leadership path. Over a period of 18 years, I transitioned from teaching in the classroom to leading science curriculum, instruction, and assessment for 60 K–12 campuses and nearly 42,000 students. I am currently the director of strategic initiatives and the principal for focus schools. I held various roles as a middle and high school science teacher, researcher, state and district science supervisor, and president of a national organization. After a brief discussion of my journey, I will reflect on my experiences through the lens of leadership themes that emerged.

CHAPTER

6

The Leadership Journey Begins

I grew up in a small town in the Midwest. Because our house was on the edge of a wooded ravine, I was always exploring. I had a deep love and fascination with the outdoors and found a world of my own design as I built forts, climbed trees, and developed sledding runs with my friends in the winter. I continued my adventures in the summer at a small cabin on a lake where I spent most of my time fishing.

Organized sports taught me important lessons. I got serious about wrestling in high school, and later, when I wrestled on the collegiate level, I had opportunities to get to know talented and dedicated team-mates. The level of competition on the team is said to raise each individual's abilities and worldview. That was definitely the case for me. Experiencing a small dose of high-level training and competition influenced how I approach any challenge in life. At a minimum, college wrestling led me to appreciate how much work I needed to invest to succeed.

My personal mastery of learning and trajectory toward a doctorate did not always come easy. Getting by with talent but poor study habits and lackluster science courses throughout my K–12 education eventually caught up with me. The toughest academic challenge happened to be in my first college science class, but instead of giving up and switching majors or losing scholarships due to bad grades, I searched out new peers who seemed to have their act together in science. After observing their habits and then modeling my habits on theirs, I succeeded in the class. I never looked back once I realized I held the power to control my academic destiny.

A Challenge

My first major leadership challenge came through working a job in residence life during college. I worked as resident adviser and director in the college dormitories, where I was responsible for anywhere from 40 to 400 students. I helped with community building, identifying struggling students (socially or academically), and enforcing dorm rules. The resident director position pushed my growth as I was given the authority to interview, hire, and appraise staff.

I entered a master's program in biochemistry right out of college. Unlike my rough academic transition from high school to college due to my own ineffectiveness, in graduate research I was applying myself but still not producing results. I was assured by my adviser that repeated failures at the lab bench were okay. Agitated, I consulted a colleague in a different research lab who was willing to listen. I remember that she lit up with an idea that led me to a totally different approach. She was right, and within a matter of days, my months of failures changed to success. I attribute this simple act of reaching out to a colleague as my first experience of being aware of the value of a professional network. I ended up completing the master's degree along with a successful publication (Blake et al. 2007).

A Bold Move

Upon completion of my master's degree, I made a bold move. Instead of immediately entering the workforce, I got married, and my wife's work led us to a rural area in Nebraska. With no research and development jobs available, I decided to change direction. I entered preservice teacher training in high school chemistry, though I felt unsure of this decision. I knew the science well, but my first experiences in relating it to students felt ineffective. Fortunately, I was placed with an accomplished cooperating teacher and warmed up to science teaching over the course of the year. My co-teacher taught me simple steps to increase my presence with learners, and I fell in love with the job.

About five years into teaching, I began to get restless. Initially, I found great satisfaction in the day-to-day challenges in middle and high school science classrooms. I worked in both a very small school and one of the largest schools in the state. I guided students to connect in-school and out-of-school learning experiences as I coached wrestling and sponsored teams such as Science Olympiad. Following the advice of a colleague, I took on a student teacher as a professional growth opportunity. I think having a student teacher, another professional in the room, pushed me to reconnect with some of my earlier interest in being a leader. Around the same time, I was becoming more acquainted with the administrative staff at the large high school. As I spent more time with an assistant principal, I felt a calling for the first time to building-level leadership. (A less altruistic reason was my desire for higher wages.)

Considering Administrative Leadership

With an eye on moving into administration, I entered a doctoral program. There were many parallels between my first experience with a bold move from laboratory to classroom and heading back for graduate study in educational administration. Like previous experiences with changing major directions in life, my negative attitude showed up. I was concerned about the additional time and money I was about to spend and impatient to reach my future earning potential. Fortunately, my pessimism melted away as I started my first class in leadership. I met others in a similar stage of life and with familiar backgrounds. I realized that I could still re-engage with my original interest of obtaining a doctorate. I also became aware that it was no longer just about the money. Spending time with a doctoral adviser who was recognized by many of the state's top educational leaders cannot be understated for the power it had on my path. Leadership is what I was about all along, and now I was choosing to embrace it.

I became motivated to learn and do more as a leader shortly after beginning graduate school. In a large district with approximately 3,000 teachers, many of whom hold administrative certificates, waiting your turn for an administrative role for years is common. When looking in the employment classified advertisements in the local paper one day, I saw an advertisement for a state science director position. I remember telling my wife about it, and we shared a laugh, thinking it was beyond my skill set. I got over my fear of rejection and sent an e-mail to the current state director. Because this became such a pivotal moment, I saved the original e-mail exchange. I share it with the hope that it encourages you to reach out to leaders in your life who are ready to lift you up, even when you question yourself.

> *It's been great working with you the past few years at State Science Olympiad running quiz bowl. However, I'm writing for a different reason. I've taught science around the state for 10 years in Class A schools (Lincoln North Star) and Class D (Loup City). I'm currently a couple years away from my doctorate at UNL in Educational Leadership, with a focus on building level administration. Then I saw that you must be retiring this year with the position advertised for the Director of Science Education. I was just curious if I could talk*

to you sometime about the requirements of the job. The description was pretty clear, I'm just curious if you think I would be a fit for that position. I attached my resume if you had a minute to look it over and give me your feedback on this idea.

The state director's response: "Your background is similar to mine when I applied for this job 27 years ago. I would be more than happy to talk with you about the job requirements, and other related issues."

During this time, I was beginning to grow a network with other leaders who were becoming mentors. I would commonly consult with them before applying for positions, especially those roles I knew little about. I share this next response as an example of someone who was well intentioned, though she almost derailed my journey: My leadership mentor at the time said I should not take the state science supervisor position. Her experience in educational administration led her to believe the role of the state agency of education was not preparation for school leadership. She worried that by spending time working at the state level, I might change my original dreams of serving in school building administration. I followed her advice and did not apply for the job. Luckily, it was not my only shot at this initial leadership role. The state position reopened a few months later, and I changed my mind.

A Position as State Science Supervisor and Expanded Opportunities

All of the energy I had invested in getting to know thousands of students over the course of 10 years was now invested fully in learning and networking in my new job as state science supervisor. With the passage of the *Next Generation Science Standards*, the voluntary national science standards, there was a stir in the air about what was to come. Without state science standards to revise yet, I had some time to read, think about my responses to stakeholders, and get to know everyone in the department. My daily tasks of just figuring out the new information were ideal for a new leader.

The state role also came with an opportunity to engage on regional, state, and national stages. I had more opportunities to connect with other Nebraska science teachers at a level I had never experienced. I began to form a larger network through many national science organizations: Council of State Science Supervisors, Council of Chief State School Officers, and National Science Teachers Association (NSTA). I also heard about other leadership groups, such as the National Science Education Leadership Association (NSELA).

I recognized and honed leadership skills in my first year out of the classroom. I discovered many small successes with my positive perceptions of others' abilities and a collaborative style that worked in the bureaucratic environment of a state agency. I learned about the challenge of coming up with a one-size-fits-all policy. I worked to develop the statewide science summative test and lay the groundwork for updating Nebraska science standards. On a personal level, I realized I could be away from students but keep the feeling of serving students and teachers close to my heart.

One key to my success was maintaining relationships with those who came before me. I could hear their thoughts, but it did not mean I had to use all of them. One good suggestion I followed was joining NSTA's Committee on Coordination and Supervision of Science Teaching. Joining a committee may seem intimidating if you have never considered participating in volunteer governance for a large organization; however, the amount of work invested in showing up to the national meeting and participating on a few calls is small compared with the reward of having a large network that connected on a deeper level than simply attending a meeting. I have continued to take this type of encouragement when trusted leaders tell me to apply for new roles. After a few more years and national committee chair roles, I finally ran for my first major position and was elected president of NSELA. I now encourage other leaders to get more involved in state and national leadership organizations as I work to pay it forward.

Moving to District Leadership

After a year of working at the state agency, I received a cryptic e-mail from a mentor—the same person who had caused me to second-guess my original intention to apply for a job at the state level. She shared that the district-level science supervisor position in my city was vacant. Little did I know that my professional network had identified me as a possible person to fill this role and had strategically worked to contact me.

I hesitated to leave the state role because I believed in helping all schools. As is the case in many states, the pay and budget at the district level were substantially higher, so I could not refuse the opportunity. I was also excited about the sustained relationships I would get to develop with one group of teachers. I did get that job as the K–12 science curriculum specialist, and the majority of my seven years outside the classroom have been spent in this and other roles in Lincoln Public Schools.

I would love to get into the details of all I have been able to accomplish over the six years in my district, but it is beyond the scope of this profile. I have updated the elementary science curriculum, added precision to secondary-science curriculum around three-dimensional teaching and learning, and learned to juggle hundreds of requests for decisions each day. It is never easy, but I have found an acceptable balance between job responsibility, family, and my health. Most important, I have been able to help other leaders grow, some in my district and others through national volunteer work. As some leaders told me early on about the long hours, I must share the best-kept secret: I find the work extremely interesting and rewarding, to the point where I do not feel like I am actually working.

Reflections on Leadership

Although the stakes are higher, moving to leadership has been a rewarding journey that has brought me closer to who I am as a person. I am going to look back on forks in the road during my journey, including childhood and various professional roles. I have organized my leadership journey through the lens of four themes that emerge: autonomy, courage work (leadership), networking, and bold moves (saying yes) to other opportunities.

Autonomy

My job is all about change, and I think having a start in life that allowed alone time for creativity led to my wild heart. To this day, I am able to maintain autonomy even when it could cause friction with others. I accomplish much of what excites people about what I do by being willing to try new approaches.

In many ways, the new state-adopted three-dimensional science standards ooze change, forcing me and others to embrace them. We are shifting from the traditional academic script many of us experienced as students ourselves to meet the innovations in *A Framework for K–12 Science Education: Practices, Crosscutting Concepts, and Core Ideas* (NRC 2012). I am sensitive to the pace while I work to introduce change. I need to hold teachers accountable for the core of the shifts and not create resistance. Some of my leadership learning comes from working on my doctoral degree. I have grown into my role through trial and error with implementing new ideas. I seek out and sustain relationships with many types of leadership mentors at all levels of the system, especially teachers. I realize that true change is slow and difficult work accomplished over many years. Although this pace has resulted in an occasional negative surprise, for the most part many teachers have joined me in enjoying successes at the district level. In thinking back to where I get my drive, my ability to sustain new ideas may relate to my past lessons from athletics.

Courage Work (Leadership)

My first intense experience with leadership was working in resident life in college. Although I have had many experiences since then, nobody could wave a magic wand and make leadership easier for me. I had to work through it myself during the transition to each new role. On many occasions, the drive to go over, under, and around a challenge came from pure stubbornness. The courage to create a vacancy on my team has been the toughest of personal challenges, but I refuse to accept mediocrity when thousands of students' futures are at stake. My formal studies for a doctorate in leadership and my willingness to listen to other leaders have been key to my success. Having a learning mindset helped set my internal drive and prepare me for the inevitable lonely or stressful moments.

Networking

During my job as a state science supervisor, I was told staying in my cubicle would not lead to a collaborative vision of science. I was encouraged to get out and network with others. Meeting with others to exchange ideas was a big part of my journey to leadership. The mission at NSELA, where I am past president, can be summarized as "advocate, collaborate, educate." As *collaborate* is one-third of our mission, I think it may be the core that allows everything else to happen. As both attendance at national volunteer meetings and the average age of members increase, I feel many younger leaders may be left behind. With a family and a busy job, it can be hard to make sacrifices to get additional learning if it involves travel. The reward of the extra miles to get to meetings has filled my bucket and prepared me for the challenge of day-to-day life leading change in a school district.

Connecting public schools to private business has become the next chapter in my leadership journey. I am once again humbled and learning through many small lessons with a great partner from the Lincoln Chamber of Commerce. We decided with others in our city to apply for inclusion into the STEM Learning Ecosystems Community of Practice—a global movement devoted to achieving dramatic improvement in how students learn by connecting in-school and out-of-school STEM-rich environments—gathering a cross-sector team that included business and industry partners, afterschool educators, policy makers, and summer program leaders to create the Lincoln STEM Ecosystem. Although the effort is still young, we were selected in 2019 for this opportunity.

Bold Moves (Saying Yes to Opportunities)

Throughout my leadership journey, there was a certain element of being willing to take risks. I was open to taking the leap involved with major transitions in life. If I had to rank the amount of thought and nervous energy across all of these changes, leaving the K–12 classroom required the most thought and effort. I was exposed to lifelong classroom teachers and idolized them. After leaving the classroom, I reconnected with my cooperating teacher from many years prior. Part of me felt like I was letting him and others down as I changed jobs. Even though he taught his whole career, he was not surprised at my job change, saying it made sense. My cooperating teacher left me with endless leadership possibilities that I did not even dream of at the time. He affirmed for me that I need to be who I am and not follow others' expectations.

Through the writing of this profile, I got to spend a few days with Rodger Bybee. Valuable time with Rodger brought a side benefit. I worked closely with him and gained candid evaluations of how I spent my time and carried myself at work. Through this interaction, I more deeply appreciate how a few words from a trusted mentor can make me a more effective, less apologetic, and more thoughtful leader.

RODGER'S INSIGHTS AND INTERPRETATIONS

I had wonderfully insightful days visiting with James and shadowing him in his work for Lincoln (Nebraska) Public Schools.

The first meeting of the day was with the curriculum director to discuss considerations of the curriculum for a new high school emphasizing medical and health sciences. During this discussion, James mostly listened. He did agree to complete several tasks. At the meeting's conclusion, the curriculum director asked James to stay and said I could stay as well to listen. This meeting offered insight as it was about a new position for which James was being considered; the curriculum director began by expressing concerns that the advancement for James may result in some conflicts with her position. I was interested to see how James responded because such a situation is not unusual for leaders and requires being sensitive to interpersonal relations, especially the other person's concerns. James listened so he would understand the situation, then responded to the fact that the curriculum director had concerns. Finally, he suggested that the new position likely would not have significant conflicts with her position and that any conflicts could be resolved.

The third meeting of the day was a brief videoconference with a colleague, an engineer at a business in Lincoln, to prepare for an upcoming presentation on women in STEM education. I noticed again that James listened well, as well as the fact that he worked with this colleague while still allowing her to lead. I also observed that his contributions to the discussion complemented his colleague's position and role in the meeting.

After lunch, we left the office for the next meeting about the Lincoln STEM Ecosystem. The STEM program focused on a working relationship between a representative of the Lincoln business community and the public schools. Each year, a significant number of students graduating from Lincoln Public Schools go directly into the workforce. This discussion centered on the question "What should the STEM Ecosystem project try to implement?" Both James and his colleague contributed ideas and potential issues with implementing the ideas in various curriculum programs. James had to respond to several suggestions about ways the Lincoln Public Schools curriculum should change, something that is consistently a challenge for leaders. Leaders such as James have numerous other components of the education system that must be considered when a colleague proposes specific changes.

Finally, we met with 12 science teachers who were invited to an informal discussion of instructional materials adopted by the district. The science teachers expressed their concerns about the programs. For example, some thought several activities were repetitive, and others did not understand how the content aligned with Nebraska's state science standards. James asked clarifying questions and made a few comments in the two-hour session. In general, he listened. I noted later that most of the concerns were within the reasonable parameters of the Concerns-Based Adoption Model. James knew of the model and agreed with my assessment.

In all meetings, James demonstrated varied abilities as an effective leader. During the day, his leadership included the knowledge and ability needed to analyze and propose a new curriculum program; an ability to listen and respond appropriately to others' sensitive personal concerns; the willingness to adapt to colleagues' suggestions; the design of an effective presentation for professional development; a recognition that district priorities included implementing new programs; a balance between recommendations for reform and priorities in the system; and an ability to respond to concerns within limits of time, budget, quality of programs, and effectiveness of instructional practices.

James's profile reveals a couple notable insights. First, his experience offers a lesson about adaptation and learning from mentors while conducting scientific research in graduate school when he listened to and learned from a colleague to make an effective change. Second, he made a transition to science teaching and demonstrated his ability to lead as a classroom teacher. With an initial awareness of the role and his abilities, he began building his skills as a leader.

CHAPTER

7

Being a Teacher Leader

K. RENAE PULLEN

K. Renae Pullen is an elementary science curriculum specialist for Caddo Parish Public Schools in Shreveport, Louisiana.

NOTE FROM RODGER

In early November 2019 I attended a meeting of the Science Teachers Association of New York State, where a colleague suggested I speak with K. Renae Pullen because she was doing great work in Louisiana. After a session where Renae presented, we had a long discussion, and to say I enjoyed the conversation and was further impressed would be a significant understatement. Renae is a knowledgeable and enthusiastic educator who began her career as an elementary school teacher (fourth grade) and has continued in several positions in Caddo Parish, Shreveport, Louisiana, for more than 20 years.

Ms. Pullen has a bachelor of arts degree in elementary education from Northwestern State University and a master's degree in educational leadership from Louisiana State University Shreveport, and she is certified as a teacher leader by the state of Louisiana.

She has received numerous awards, including the Wal-Mart Local Teacher of the Year and the Presidential Award for Excellence in Mathematics and Science Teaching.

Teaching Was Not for Me—Until . . .

Many of my colleagues share that they have always wanted to be a teacher. That is not my story, even though I come from a family of educators. In the 1960s, my grandfather was a principal in the Sweetwater, Texas, school district for Black students. When schools integrated, he was the second Black principal appointed to an integrated school in Sweetwater. My mom was a kindergarten teacher and went on to become an elementary school principal. My grandmother, uncle, and aunt were all educators. As a child, I knew one thing: I did not want to be a teacher.

As a family of educators, we believed in the importance of learning. We believed, and still believe, that learning is essential to achieving success and finding opportunities. Knowledge and critical thinking are essential for us to make good decisions, understand the world around us, and not be taken advantage of. Indeed, learning was meaningful to me, but I wanted to do other things. After all, one cannot become rich from teaching. I did not think teaching was for me, and I believed that until the summer I fell in love.

During my junior year of high school, a teacher shared an opportunity to engage in community service and be part of a summer program she was sponsoring called Rainbow Magic. I volunteered

mainly because I did not want to get a summer job. In Rainbow Magic, we would work with local children in a community literacy program. We were paired with one student with whom we would build a relationship and provide community-based learning experiences while exploring literacy in exciting ways. Our goal was to work with our young students so they could improve their literacy skills and be more successful in school.

On the first day, a little white van brought our new friends to us. I was assigned my friend, and the first thing he grumpily told me was, "I don't even want to be here. I want to be at home." I responded, "Me too, but my mom made me come. Look, we're here, so let's have some fun today!" He giggled, and we were on our way to being fast friends. It was an amazing experience. We visited local parks and museums. We took a day trip to a zoo in Texas. We performed plays based on the silly things we had written. We read about all sorts of things, and I discovered that my friend loved to read about animals, space, and interesting people. He had never really had an opportunity to read nonfiction before. I also discovered that he struggled with digraphs (a pair of letters that in speech present a single sound, such as the *ea* in *beat*). It was so much fun finding things to read with my friend and watching him explore new things. As the summer closed, he understood digraphs and learned so much about our natural world, art, and history. I was incredibly proud of him and his growth. As my friend climbed into the little white van for the last time, my heart was sad but full. I really could not deny it anymore. That was what I wanted to do all the time. I had fallen in love with teaching.

A Beginning

After I completed my undergraduate work, I began teaching at a rural school in Louisiana as a self-contained third-grade teacher. Even though I was encouraged to focus on English language arts and math, I made sure I taught science and social studies every day. I found that I could get my students to engage in literacy more if we wrote and read in science and social studies. I expected my students to work with multiple types of text and communicate in a variety of ways. My students enjoyed engaging in science investigations, reading about science, collaborating with one another, and demonstrating what they learned through text and graphic forms such as charts and graphs. I would like to report that I was an extremely effective English language arts teacher, but I am not certain that was true. As a first-year teacher, I did not have the language arts teacher tool bag that my veteran colleagues had. What I did have was an expectation that my students would engage in literacy often and in various ways. And it worked.

An Opportunity

After my first year of teaching, my principal told me about a professional development opportunity that my district was hosting called the Elementary Science Program (ESP). She wanted to submit my name to be a part of the program because I enjoyed teaching science. The program was an intense, three-week professional development summer institute that provided a deep dive into content knowledge and effective evidence-based pedagogical practices of science instruction. The summer institute was followed with several collaborations during the school year. ESP gave me an opportunity to build my content knowledge, hone my skills as an elementary science educator, and support my professional learning with research-based approaches. I had never participated in such a robust professional learning experience. The program gave me the pedagogical content knowledge necessary to make me confident as a teacher of elementary school science. It also gave me the opportunity to collaborate with a local university professor and district colleagues.

ESP empowered me as a teacher. I learned that it was all right not to have all of the answers. I applied the new techniques and pedagogical content knowledge to my classroom the following year and shared what I learned with my colleagues. In our collaborative meetings, we talked about what worked, what did not work, and what we would do next for our students. I had the least experience in our cohort, and I was unaware just how much my colleagues struggled to have the time to put what we had learned into practice. My school administrator supported me in teaching English language arts and mathematics as well as science and social studies every day, but I was shocked to learn that was not the norm. When I read some of the books from our program, I was even more disappointed to learn that other teachers around the country lacked similar support. I did not know what I could do about it, but I knew two things for sure: I was going to teach science in my third-grade classroom every single day, and I still had a lot to learn.

More Opportunities and a Mentor

Over the next few years, I sought professional learning and service opportunities. I worked on school and district committees. I was lucky to have school administrators and district science supervisors who valued my voice and empowered me to engage in leadership opportunities. At the time, we had elementary, middle, and high school science supervisors. All three were supportive of me and my work with students in the classroom. They encouraged me to write grants, seek professional service within and outside the district, and work on my master's degree. My elementary science supervisor, Dr. Patsy Latin, was incredible. She conducted exciting professional development, was extremely supportive, and loved to see science learning and student collaboration in the classroom. She took chances and had many innovative ideas when it came to technology and student collaboration. It was invaluable for me to see a Black woman leading science education effectively in my district.

Dr. Latin encouraged me to share my expertise and showed me ways I could get funds to support my classroom programs. I began conducting professional development for colleagues in my school and district. After four years of teaching third grade, I was asked to move to fourth grade, the high-stakes testing grade, to teach science and social studies. I was excited to focus on my two favorite subjects. My principal stated that she expected me to continue having my students do mathematics and literacy in my science and social studies classes, which was perfect because that was exactly what I wanted to do.

Over the next few years, I continued to grow as a professional. I wrote grants for resources and trade books to teach science. I started to speak out more about the disparity between time and resources for science and social studies compared with time and resources for English language arts and math. I participated in a yearlong book study in which we read and discussed several books about learning and cognition, an amazingly innovative professional learning opportunity that was teacher-designed, teacher-led, and collaborative. This study helped me think more deeply about how I could change my teaching and

assessment practices to improve students' outcomes, and it taught me that I had the capacity to read and understand educational research to improve my practice. After a few years, I transferred from my rural school to teach fourth-grade science and social studies at a school closer to home. After my first year at my new school, I was selected as teacher of the year and later as the elementary teacher of the year for my district. A couple years later, I began to work on my master's degree in educational leadership.

Graduate Education and Another Mentor

As I worked on my graduate studies, I discovered that administration was not my passion. I wanted to lead, but I did not want to lead by being an administrator of a school. Fortunately, I had professors who showed me I could think of leadership differently. One of my undergraduate science education professors, Dr. Linda Easley-Tidwell, was part of my graduate program faculty and, like Dr. Latin, a science educator I aspired to be like. Her influence was apparent in my classroom. She strongly believed in science inquiry and student collaboration. She had high expectations and was a generous teacher. Furthermore, she had a wicked sense of humor. Dr. Easley-Tidwell taught me to encourage students' curiosity, explore their questions, and learn how to capitalize on students' collaboration. She loved when we asked questions and seemed genuinely excited to explore them, and her classes were both tough and fun.

Dr. Easley-Tidwell was adamant about focusing on evidence and using research to inform instructional and leadership decisions while understanding the limitations of research. She encouraged me to lead more from my classroom and begin presenting more at the state and national levels to share my expertise. I was hesitant, but I began to welcome science methods preservice teachers into my classroom because of her encouragement. She would often talk about an article or book she read, then I would read those resources. I was incredibly fortunate to have such an amazing mentor guide me through my undergraduate and graduate studies for teaching and leadership. After completing my graduate studies, I decided that the only leadership position I wanted would be related to instruction and teacher support, specifically science instruction and teachers' professional learning.

Recognition as a Leader

During my graduate program's final year, I was nominated for the prestigious Presidential Award for Excellence in Mathematics and Science Teaching (PAEMST), the highest honor bestowed upon a math or science teacher, and received the award for science teaching in 2008. I also completed a teacher leadership program sponsored by the Louisiana Department of Education and a partnering university. During my teacher leadership training, I learned about elevating the profession, sharing my teacher voice, building strong administrator-teacher relationships, and leveraging resources from professional organizations to continue growing professionally. I started to seek more leadership opportunities, such as serving as grade-level chair for my fourth-grade team, participating in school- and district-level committees, pursuing professional learning opportunities, and advocating for more time and resources to teach science in elementary schools. I began to get involved with the Louisiana Science Teachers Association. I wanted to learn more about creativity and STEM, so I applied for a Fund for Teachers fellowship to study creativity in STEM in Barcelona, Spain, and for a National Endowment for the Humanities fellowship to study the American skyscraper in Chicago, hoping to get at least one opportunity. I was shocked when I was awarded both. These experiences taught me to be a courageous professional.

It can be overwhelming to put yourself out there professionally. Not being awarded a grant or not being placed on a committee where you think you can be an asset can feel like you have failed, but I learned that I had to take those chances for my students and myself. How can I expect my students to be brave if I am not?

Yet More Opportunities

After I received the PAEMST, more opportunities opened up for me, including an opportunity to speak at a national meeting with fellow presidential awardees. It was my first national meeting of that kind, and I was excited and nervous to be in that room with so many accomplished educators. We talked about the importance of science and math education and how we could leverage our collective expertise to better prepare and support teachers in STEM subjects. At one point, I became frustrated because the conversations seemed to be primarily about elementary math and not science. It always seemed that when I was a part of these conversations in other meetings, elementary educators were included, but the discussions never addressed elementary science learning. I decided to speak up for elementary science because if we were not making it a priority in this national meeting with like-minded professionals, we could not expect elementary science to be a priority on the state and district levels. It was also important to me that our discussions positioned elementary teachers, who are generalists, as capable teachers of science who have specific expertise that can be leveraged, rather than having a deficit view of elementary educators as teachers who are not subject-matter experts. My voice was heard and acknowledged.

At the time, I did not know that there was a representative in that meeting from the Teacher Advisory Council for the National Academies of Sciences, Engineering, and Medicine (NASEM). A couple years later, I was contacted about being a member of NASEM's Teacher Advisory Council. I joined the council, whose purpose was to "increase the usefulness, relevance, and communication of research to educational practice; help the research community develop new research that is informed by practice; provide advice about how other National Academies' programs, initiatives, and recommendations can be most effectively implemented in schools; and offer guidance about how the National Academies can best communicate with the teaching community in the United States." As a member of the Teacher Advisory Council, I had the opportunity to attend convocations, collaborate with researchers and fellow practitioners, and share my expertise and wisdom from practice in meaningful ways. In 2018, I was asked to be a member of the Board on Science Education (BOSE) for NASEM. BOSE aims to investigate how science is learned by everyone from young children to adults and to explore evidence that uncovers how science can be taught effectively in both formal and informal settings.

Leaving the Classroom

After completing my graduate work, I was certain there was no position that would allow me to remain in the classroom. In 2014, I was encouraged by my district leadership to apply for an elementary science specialist position. My chief academic officer revised the curriculum supervisor positions to be called

curriculum specialists so that specialists could focus on supporting teachers. I was torn about applying because I was not ready to leave the classroom, but this was the only administrative job that would inspire me to leave. I would have the opportunity to be to other teachers what Dr. Latin had been for me and did not want to miss that opportunity. Since becoming a science specialist for my district, I have been fortunate to be able to continue leveraging my voice and leadership. After all these years, I still seek opportunities to grow professionally. I still champion for more time, resources, and support for elementary science instruction. And I am still in love with teaching.

What I Have Learned as a Leader

Throughout my more than 20 years as a teacher and content specialist, I have had many opportunities to lead. For 16 years, I led as a classroom teacher, and I still believe in teacher leadership—I do not think teachers have to leave the classroom to lead. Listening to teachers' voices is imperative, and teachers' expertise should be leveraged and elevated. Teachers should not be afraid to get involved with decisions that concern instruction, assessment, and policies. Over the years, I have learned that building my capacity and developing connections with colleagues strengthens my voice. I engaged with state and national science teacher organizations and collaborated with colleagues in person and on social media. I was relentless as I shared my experiences and expertise. I read research to inform my teaching practices and learned to prioritize my professional service so that I only sought opportunities that would help me achieve my professional goals. I helped improve outcomes for my students and made it difficult for others to manipulate me into engaging in problematic and inequitable practices because I was familiar with what research suggested.

As I reflect on my experience as a learner, teacher, and leader, I have uncovered three truths that continue to guide me as a leader:

1. Leaders are caring. I do not see leadership as a structure of position and power, but instead as a service that is influential and visionary and can empower others. It is important to me that my leadership is rooted in generosity, equity, and empathy. I freely and kindly share my knowledge and expertise and want to leverage the assets my colleagues bring to the table rather than focus on what some may claim as a weakness. This is especially important to me as an elementary science educator who has often been dismissed as a generalist who is not a subject-matter expert. Caring leaders seek to acknowledge their biases and are aware of their impact on others. They understand that their colleagues' experiences and voices have value and should be appreciated, used, and respected. I want to support my colleagues and honor what they have to contribute. Genuinely caring about others' work and well-being is an essential quality for good leaders.

2. Leaders are learners. As I started my leadership journey, I was fortunate to realize that while I had good pedagogical content knowledge and an interest in building my capacity to lead, I still had so much to learn. I knew I did not have all of the answers, so I asked questions. I sought opportunities to collaborate with others who had different knowledge, experiences, and expertise. Investing in my own growth and development as a science content leader became a prior-

ity and continues to be one. Knowing where to go to build your capacity as a content leader is crucial. Professional organizations like the National Science Teaching Association and state science teaching organizations allow you to engage in professional learning opportunities, discover outstanding books and journals to build capacity, and collaborate with colleagues. I find that collaborating with and learning from education practitioners and researchers have made me a better, more informed educator and have actually made being an elementary science leader feel less lonely. Reflecting on opportunities to build your capacity as a leader through professional learning experiences and collaborations may be uncomfortable, but it is essential. Professional growth and development are assets that are worth the time and resources.

3. Leaders are brave. I am unafraid of being vulnerable as a leader. I acknowledge that I am not a sage. Therefore, I am dedicated to learning and growing professionally so I can achieve my goals. Improving science education and elevating the profession are my moral imperatives. To achieve these goals, I decided—while I was still in the classroom—that I needed to take risks not only for myself but also for my students. I applied for grants and fellowships, spoke up for science time and resources in elementary schools, engaged in difficult conversations, shared my expertise, and sought community partnerships. I was scared at first because I could have been rejected or failed at achieving a goal. Occasionally, I was rejected and did fail at achieving goals. Luckily, I had awesome colleagues and mentors who encouraged me to be reflective and refine my approach. I could not improve my craft if I was unwilling to be courageous as a leader. Dr. Martin Luther King Jr. (1963, p. 123) said, "Courage faces fear and thereby masters it." My failures were learning opportunities that informed my future successes. Fear and uncertainty can be powerful disincentives, and they should not drive you as a leader. Brave leaders take chances, engage in challenging conversations, and embrace diversity of thought.

Conclusion

I have been fortunate to have worked with amazing leaders who embodied these truths and were great mentors. From my professor who taught me as an undergraduate and graduate student to my district supervisors and school leaders, these leaders encouraged me and supported my endeavors to improve the quality of instruction so my students could succeed thanks to my leadership as a classroom teacher and district leader. These leaders affirmed my credibility as a teacher and leader. I know that it is sometimes difficult for teachers to feel empowered to share their voice, but I believe it can be detrimental to student growth when teachers' voices are suppressed and their experience and expertise are not valued. Teachers are experienced with building relationships between students and families. They are the ones tasked with doing the work of instructing students every day. It is a woeful mistake for administrators and even policy makers not to take advantage of the experience and insight that teacher leaders provide. Administrators and policy makers must make the courageous decision to learn how to empower teachers and leverage their talents and abilities. Leaders should encourage teachers to take risks, apply for grants and fellowships, grow professionally, and take a seat at the decision-making table.

As I think back to the young girl who fell in love with teaching one summer, I am proud of the teacher and leader I have become. I still have work to do, and I am excited to continue to hone my craft and have a positive impact on the profession. I was wrong about teaching not making me rich. When I look back on my students' successes and the impact that I have had in my career, I realize I have become rich in ways I did not expect.

RODGER'S INSIGHTS AND INTERPRETATIONS

Renae's profile is quite insightful, especially for a midcareer science educator. Please note several features of her profile. First, as a beginning third-grade teacher, she understood the role of leadership in terms of her classroom and students. Second, she sought opportunities to develop as a better science (and math, literacy, and social sciences) teacher. Finally, beginning with her own learning and leadership, she progressed to working with colleagues in professional learning communities and eventually contributed at the district, state, and national levels as a leader.

Renae's progress can be seen in the evolution of both her knowledge and her skills in science and teaching, as well as in her understanding of leadership and her improved abilities. The fact that she recognized and followed modeled behaviors and the advice of mentors is an important insight that others can apply to their own leadership. She also sought and used high-quality instructional materials, which can be invaluable for any teacher or leader, particularly those early in their careers.

Renae's three truths about leadership convey further insights that all leaders should keep in mind: Leaders are caring, leaders are learners, and leaders are brave.

CHAPTER

8

My Leadership Journey

CYNTHIA LONG

Cynthia Long is the K–12 science curriculum coordinator for the School District of Osceola County in Kissimmee, Florida.

NOTE FROM RODGER

Cynthia Long has taught K–12 students and teachers in English language arts, mathematics, and science in addition to writing textbooks and doing scientific research. She continues to support teachers in the development of high-quality curriculum materials and instructional strategies across disciplines. She has served as an assistant principal at a STEM high school in a high-needs area in Denver, Colorado.

Cyndi has a bachelor's degree in molecular, cellular, and developmental biology from the University of Colorado Boulder and a master's of education in curriculum and instruction for secondary science from the University of Washington in Seattle.

My Journey

As I write this reflection, I find myself listing things I've done throughout my more than 30 years in science education like a checklist of tasks. But this list did not get to the depth that I wanted, which would describe all that went into decisions and directions, learning and leading. My circuitous route to where I am today formed who I am as a leader. The people, experiences, and interactions—both positive and challenging—gave me tools, confidence, trust, and passion to do the work I do.

Early Decisions

As the youngest of four, I was known as the last Meis (my maiden name). I tended to be fairly adventurous. I played sports and did fine in school, always liking science the best. I gravitated toward friendships with multiple groups rather than sticking with one. I liked to voice my ideas but did not like the spotlight. When it was time to decide on a career, I leaned toward science but did not know what options were available, so I chose nursing. During a required visit with the high school counselor, I shared my plans. He asked, "Why nursing?" As I explained the cliché answer, "I like to help people" and added, "And I love science," he said I really should become a doctor. Honestly, that thought had never crossed my mind, but after that I couldn't think about doing anything else. I attended the University of Colorado Boulder (CU) and majored in what I thought all pre-meds majored in: molecular, cellular, and developmental biology.

Contributing Research

After graduating from CU and grappling with whether to pursue medical school after being waitlisted, I got a job as a researcher at the University of Colorado Health Sciences Center. I loved the work. Although I learned molecular biology at CU, I really came to understand it in the lab. I worked with brilliant scientists who were studying the complexity of cancer, then with others researching HIV. Doing bench research when HIV was a novel virus gave me the opportunity to experience the impact such work could have. The day-to-day research didn't seem important at the time, but looking back, I'm filled with appreciation for the chance to work on and contribute to this research. With guidance, I asked questions and designed experiments that would help answer questions such as "How can we target infected cells and destroy them?" and "What conditions are best to replicate the DNA of infected cells for study using polymerase chain reaction (PCR)?" I'd collect and analyze data and share my findings with colleagues. Being responsible for research that would advance an understanding of HIV and be used to explore treatments inspired me to delve deeper into science. I was honored to be on the writing team for several published papers. I was a part of a group with members who mentored and trusted me. Science research taught me that even though much of the time an experiment *fails*, it is still always a success because something is learned. The experiment leads to more questions and more explorations. Science is about curiosity and discovery, asking questions and getting more information to help answer questions. While working in this position, I was asked about participating in a side project, which was the first time I had the opportunity to plan and facilitate a professional learning event for teachers. This opportunity was my springboard into science education.

A New Opportunity in Scientific Research

In my personal life, this was also about the time I got married and moved to Seattle, Washington. I was ready for a change and was hired for as a research scientist at the University of Washington (UW) in Seattle. This change was a new adventure—a new risk—but an exciting one I was ready to embrace as we loaded our two large dogs in our car and headed to the Pacific Northwest.

As I settled in as the lab manager and research scientist in the Department of Bioengineering at UW, I had an incredible opportunity to work with a new professor. Performing tasks such as setting up a lab in a non-lab space and introducing graduate students to the work was a great fit for me. The graduate students all brought unique backgrounds, experiences, talents, quirks, and interests. I was privileged to work in this environment, being surrounded by people from all over the world and watching them grow. I also watched how my boss mentored and taught these students. My boss put me in the cycle to present my research at lab meetings. Scary! I dove in, shaky voice and all. It was awful at first, but the more I did it, the more comfortable I felt. My boss included me, respected my work, and expected me to present. What a gift—without his insistence, I never would have found my voice. He also introduced me to the idea of educational outreach, and before I knew it, I was jumping into that, too.

Educational Outreach

My first and most impactful program was one my boss connected me with and asked me to lead. That boss was, and continues to be, a connector and an advocate of educational outreach. This program opened up a whole new world to me as my boss gave me the opportunity to develop and teach in the DO-IT (Disabilities, Opportunities, Internetworking, and Technology) program. DO-IT empowers people with disabilities, and our work focused on technology, education, and laboratory research.

This incredible program brought a group of high school students to UW and to my lab. I designed genetic engineering experiments that gave students real lab experiences, like using pipettes, extracting and loading DNA into a gel for agarose gel electrophoresis, growing bacteria, and doing cell cultures. I worked with these students more than 20 years ago and still remember the experience vividly. Two particular moments come to mind. The first involved a young woman in our group who was blind. She stood in the back during discussions and demonstrations. When it was time for each student to load the gel, she did not come forward. I asked her why and she was hesitant. She shared that she did not think she could do this particular task, but I encouraged her to try and initially guided her hands. Then she took over, using touch to manipulate a pipette and load a gel with DNA. The pride I saw in her face and her grin will stay with me forever.

Another memorable moment came during a conversation about bioethics. During an in-depth conversation, a young man spoke about how if certain technologies existed early on in his life, things might be different for him. His comments brought theoretical words from a textbook to life. The DO-IT program aimed to create an environment in which every student could do science, contribute, and be an essential part of the experience. This group of young people gave me something that is the fabric of who I am—the belief that I can help create opportunities that might have an impact on others. I take them with me in everything I do.

After my involvement with DO-IT, I wanted to learn all I could about education outreach through every opportunity offered to me and those I sought on my own. I met everyone in every program I could find at UW who was doing any kind of science outreach. UW was rich with opportunity and expertise, and I learned so much. I spoke with and ultimately worked with scientists who ran the Human Genome Project. The scientists not only did cutting-edge research but also knew the importance of reaching out to and sharing with educators and students.

A Graduate Degree

With an undergraduate degree in molecular, cellular, and developmental biology, my love for science, and my plan for medical school not panning out, I found my way to and embraced science research. After broadening my experience in education outreach, I wanted to earn a degree in education. While I worked full time as a researcher at UW and had two children, I also went to graduate school and got a master's degree in secondary science education. I remember having thoughtful discussions with my professors about science being the center of all education. It was quite the debate, as I was passionate about science being the connector of all disciplines. I still feel that way!

After earning my graduate degree, I moved from being a research scientist in bioengineering to working as an educational outreach coordinator at the University of Washington Engineered Biomaterials

(UWEB). The goal of UWEB's educational outreach was to bring scientists, educators, and students together. Developing several programs and then coordinating and teaching them was exhilarating. I worked with bioengineers and middle school science teachers to develop a curriculum in which students would build prosthetics using everyday materials (e.g., straws, rubber bands, string). Students would study the form and function of a hand, for example, then create a prosthetic hand using these materials. Students gained background knowledge in multiple ways, then applied it to their design and model. Their designs and confidence in showing the different parts they modeled was a great demonstration of how this type of science education empowers students to be scientists.

After UWEB, I transitioned to the Fred Hutchinson Cancer Research Center (FHCRC). The education and training department and its director exposed me to new programs. I was part of a team looking at the effectiveness of program implementation and gained more experience working with scientists, science writers, educators, and students. When I think about my decade in Seattle and try to describe how I developed as a person, it is difficult to find words. I keep coming back to two: *learning* and *sharing*. Many people trusted me to build programs and find ways to connect educators and students with science and scientists. Writing this reflection, I realize how trust has influenced my work today as a leader. The other core idea that drives all of my work in leadership and instruction is to always keep students at the center and make learning meaningful. Students should have access to what scientists are doing and understand its relevance to their lives. They should picture themselves as scientists and have opportunities to be scientists and *do* science.

It's cliché to say, but I was a sponge while living in Seattle. This new world of connecting education and science showed me a path I didn't know existed. I can't say enough how much I learned from all the people I worked with at UW, FHCRC, and the surrounding school districts. The bioengineering professor at UW who gave me the opportunity to design and run my very first outreach program taught me the impact I can make on others, which had a huge impact on me. I also learned how collaboration is key.

Joining BSCS and Advancing as a Leader

Next, we moved back to Colorado, and BSCS (Biological Sciences Curriculum Study) was on my radar as a cutting-edge, high-quality nonprofit curriculum development organization. I applied for a position as a curriculum and professional development coordinator working on a new project, BSCS Science: An Integrated Approach. I had strengthened my ability to build relationships among scientists, educators, and communities during my time in Seattle. Little did I know at the time that my learning and leading would expand exponentially thanks to the mentorship of the leadership team at BSCS.

I would typically arrive at BSCS early in the morning, which turned out to be a special time because I could get to know, talk with, and learn from two people who define science education and professional development for me—Dr. Rodger Bybee and Dr. Susan Loucks-Horsley.

This part is difficult to write: Due to an accident, we lost Susan too soon. I imagine all of the conversations we could have had. She guided me and gave me a true understanding of what it means to support teachers. I remember sitting across from her and listening to her describe creating an experience for teachers as being like weaving an elegant basket: One must put in elements that are meaningful, helpful, carefully chosen, and just right for each and every person. She spoke with such care and clarity. She also taught me how change takes time. As much as we plan, we have to allow people to go through their own processing before they can be on board. I saw in her what I wanted to be like as a leader: insightful and intelligent, with the ability to apply and share knowledge in any circumstance and to emanate kindness.

Rodger truly taught me science education and how to share the dynamic and ever-evolving expanse of it with educators and students so they can experience education with curiosity, passion, and wonder. I had science research experience and had developed and facilitated professional learning and outreach to students and educators, but I still felt like I had so much more to learn and experience. With Rodger's guidance, we created a curriculum for classrooms across the country, worked with teachers in different states with diverse demographics, and saw the curriculum come to life in the hands of master teachers.

Writing a formal curriculum was new for me. I was in my office one early morning when Rodger stopped by and filled me with awe as he spoke about the biodiversity of life and how it came to be. I asked questions, and he asked them back. He guided me through inquiry and discovery, which I only just now realize. On another occasion, he asked me how the writing was going. I was stuck—really struggling with a chapter. He told me just to start writing, to put the words down and not worry about writing it perfectly the first time. Just write. Then, step away for a while, and when you come back, read it through and start to arrange and edit. Since then, even now, his words are in my head and that is how I write. I learned how to write, encourage connections, and teach from Rodger.

I have never been a lecturer. When students and teachers ask me questions, I often answer with a question. We explore answers together. Ownership of the learning belongs to the learner. And for a learner to own it, they have to be able to express their ideas and ask questions; if something isn't quite right, they need to try again and again. In all the work I've done since learning from Rodger, I have used this approach and the 5E Instructional Model (Engage, Explore, Explain, Elaborate, and Evaluate) to guide my work. I also learned from mistakes I've made. Coordinating Professional Development Week for educators from around the country was a great responsibility and taught me how to facilitate large-scale events. However, I did not vet one of the presenters before his session, and he displayed a very upsetting slide. I will never forget my regret and how the slide upset some of my teachers. Now, I always preview external presentations. I embraced leadership opportunities to the fullest at BSCS by learning from my mistakes, my colleagues, and science education leaders around the country. It remains a foundation of my work today.

And Now—a Classroom Teacher

After working at BSCS for several years and thinking about my growing boys, I felt it was time to stay closer to home. After 14 years in science education, but never in a formal classroom setting, I became a high school science teacher. Now it was my turn to implement all that I'd shared with others. I taught several science subjects and all high school grade levels. It was a blast! I felt like I was playing every day I went to work. My biggest goal? Getting students to go from saying, "I am bad at science and I hate science" to never missing a class, asking questions, and doing science. I was making the curriculum come to life, which was what I always wanted the teachers I worked with to do. Since leaving the classroom,

I always picture my students as the reason we do what we do. I see their faces, abilities, strengths, challenges, distractions, focus, excitement, sadness, and goals.

New Opportunities

Another decade passed, and I moved to an educational research organization. I continued to do professional development, mostly with the *Next Generation Science Standards* as well as some work with the *Common Core State Standards in Mathematics and English Language Arts*. I assisted in grant writing and contributed to systemic professional development. I took the lead on many projects and put together my experiences to create a story that helped guide educators. In this position, I gained an understanding of grant writing. As funding ended, I turned the page to another chapter.

Looking for something different, loving technology, and always having an affinity for the ocean, I heard about a start-up company that sounded like a great opportunity. I went to work as the science education director and chief operating officer at Ocean First Education. The initial work was identifying a shared vision: What should digital marine science education look like, and how could it be supported and implemented in K–12 classrooms? What should students be learning and doing? What should our team look like to accomplish what we want? We formed a team of scientists, educators, businesspeople, ocean advocates, and information technology and coding experts. We met, brainstormed ideas, learned HTML5 coding (which I ended up loving), watched students interact with the curriculum, revised, and kept going. I learned a lot about management, communication, and organization in this role.

New Roles: Instructional Coach and Assistant Principal

As curriculum development ended and it was time for distribution and sales, I knew this was not a strength of mine, and it was time to go back to my passion for education and working with teachers. I was hired as an instructional coach at a K–12 school. This had to be one of my favorite roles, as I envisioned, provided a plan for, and supported all secondary teachers. We were working toward becoming an International Baccalaureate (IB) Middle Years Programme school. I loved pulling great ideas out of even the most resistant teachers and guiding them toward teaching in a whole new way. We learned and navigated IB together while improving instruction along the way.

After almost two years, the superintendent asked me to move to another high school and become the assistant principal (AP). It was difficult to leave my coaching position because we had made such great progress as a school. However, the school the superintendent wanted me to move to was a STEM high school where I could contribute as an instructional AP. I developed great relationships with the teachers and students at this school and provided similar support through the perspective of an AP.

Back to a School District

My current role is as the K–12 science coordinator for the School District of Osceola County, Florida. I feel privileged to work with a team that is passionate about science education. Our goals are to provide the best learning experiences, opportunities, and pathways for students to become lifelong scientists and for our educators to love teaching science and connecting with students.

As a school district leadership team member, I can be in a meeting and listen, watch, and see ideas form, all while looking at a particular direction the district wants to take. Leadership happened over time for me. By learning every step of the way and having an open mind, I've developed the insight and experience to provide guidance in decision-making, including by trusting others and letting go so they are elevated to own their work, bring their ideas forward, and see them through. Being reflective and knowing when to speak and when to let others take the lead is a balance. Practicing helps me understand different perspectives and their value.

Reflections on Leadership

When thinking about the leadership abilities I've developed, I would make the following recommendations to guide others as they venture into leadership:

1. Be accessible. Develop a positive and open relationship. Being accessible means your team will always come to you to ask questions, brainstorm, run ideas past you, and express concerns. This open communication builds momentum toward a common goal. Always make the time to be accessible.

2. Recognize and value all perspectives and roles. In every work environment and on every team, people have different roles and perspectives. Everyone has a variety of skills, knowledge, backgrounds, and perspectives. Whether they are administrative assistants or superintendents, value what others bring to the work. Show that you recognize and value their contributions.

3. Collaborate. In leadership, collaboration is a key to successful development and implementation of any idea. When recognizing and valuing all perspectives and roles, it is important to create an environment where others feel comfortable contributing and discussing ideas. It is the leader's role to facilitate this conversation, listen well, and make decisions after reviewing all ideas. A leader should support decisions based on evidence and make sure there is a shared understanding of the direction and reasoning behind it.

4. Do your homework. Especially when new in a leadership role, it is important to take the time to listen to others, ask questions, and learn about not only the role but also the whole system (e.g., school, district, organization). Rather than immediately taking action, spend time listening to the dynamics, background, and history. This will give others the chance to get to know you and allow you to show respect for what has come before and learn information that will guide and align your direction with that of the team you will lead.

5. Choose your moments—quality over quantity. Develop an awareness of when to listen and when to share knowledge or ideas. The quality of what you share is much more important than the quantity. Listen, then be confident when you do speak. Be intentional, clear, and concise.

6. Trust and delegate. As you assume a leadership role, it is important to understand that you are a teacher. Allow your team to spread their wings and grow. Although it may be difficult at

times, let go of things that used to be part of your role and allow others to take those on those tasks. When things are done differently, if they are aligned to the goal, trust that the team has the skills and knowledge you've shared to take on projects and succeed.

7. Celebrate. Part of being a leader is celebrating your team's and individuals' successes. Elevate the contributions of others sincerely and authentically. This is not a dog-and-pony show but a sincere expression of appreciation and celebration of efforts. This celebration can take the form of a passing comment, an e-mail, or an announcement. Although this may be difficult, always try to celebrate hard work, passion, and contributions.

8. Always be a learner. Inherent in being a good leader is having an open mind and continuously learning. I've always felt that the day I stop learning is the day I need to stop being an educator. Learn from others, from situations, from successes, and from challenges—but never stop learning. We have certain expertise, but we are never the only experts. Enjoy listening to what others have learned, and grab onto opportunities to learn more. Embrace being a leader, share what you've learned with others, and allow them to grow and flourish as valued members of your team.

Conclusion

It has been a journey in itself remembering all of my mentors and how much I have grown as a person and a leader because of their trust and guidance. Looking back at lessons I've learned, through both positive and challenging experiences, has been invaluable. Imprints of significant moments, conversations, and experiences have come to light. All of my roles have been pieces of a complicated puzzle that builds a full picture of what science education means to me. It has always been integral to my beliefs to not only share with others but also influence how they learn, act, and feel. I appreciate and value others' expertise and am proud to see them shine. Navigating different roles has taught me to look for a common goal, review systems as a whole, break down the details to achieve that goal, support the work, and always keep students at the center. Although I don't know where my journey will end, the most important thing to remember is that it never *really* ends. There is always something new to experience, learn, and share.

RODGER'S INSIGHTS AND INTERPRETATIONS

Cyndi did not take the usual path of teacher leadership; she did not become a teacher and then move out of the classroom to other positions of leadership.

I gained a few insights from Cyndi's profile. First, Cyndi was an active learner in all of her new situations. Second, she seized opportunities and identified mentors in her positions. Third, she internalized lessons from these mentors while adapting those lessons to her particular situation. Fourth, early in her career, she learned to respect the dignity and integrity of those she leads. Finally, she established and maintained a larger perspective within her position and the context of the program she led.

The guidance she describes for other leaders is grounded in respect for others, respect for herself as a leader, and understanding the situation when she is poised to lead.

Leadership for STEM Education

PHILIP JOHNSTON

Philip Johnston is the STEM coordinator at Central High School in Grand Junction, Colorado.

NOTE FROM RODGER

I first met Philip Johnston in a professional seminar on leadership for STEM educators. After several discussions, I realized that he was not only a STEM coordinator for a large high school but also a licensed principal who is qualified in career and technical education (CTE). Philip has a bachelor's degree in information networking and telecommunications and a master's degree in organizational leadership, both from Fort Hays State University in Hays, Kansas.

Introduction

To understand one's capacity to lead, we must consider how the person's experiences and influences contributed to his or her knowledge, ability, and style. To say there are many ideas on leadership and how to develop is an understatement. Identifying what leadership looks like in a STEM educational environment that is inherently chaotic and messy is not simple. What I can offer are some thoughts and reflections based on my leadership for the STEM program and vision at Central High School (CHS) in Grand Junction, Colorado.

Growing a STEM Culture in the Learning Environment

After eight years of work building a national identity for STEM education and my experiences on numerous campuses across the country, it is clear to me that cultivating a STEM program is not as simple as implementing one course or school curriculum. While courses and curricula are necessary building blocks of STEM education, the real foundation must be built through leadership and culture. For instance, two schools in one state or district may have the same STEM program, but the look and feel of each program will be derived from the leadership and culture of each learning community. The overarching purpose of STEM education is to apply the design process to generate solutions to real or perceived problems, and each school has its own set of variables to consider when developing a program.

Changing our CTE program from addressing precision trades to focusing on STEM has been a yearslong process, primarily due to the amount of time it takes for the learning community to develop an understanding of what STEM education looks like in practice. How do we as leaders infuse a common understanding of STEM education across our campus culture while allowing professionals the opportunity to operate to pursue this end goal? How do we make STEM sustainable? How do we allow teachers

the autonomy they have earned through their professional qualifications and still operate as a campus with a shared STEM vision? Here are my story and answers to these questions.

A few years into my classroom experiments with STEM education, and having experienced the success students were having with career and college pathways, attendance, low discipline issues, school buy-in, and pride in being a teacher at Central High School, I began looking for a way to develop a common STEM framework for other teachers to apply. In 2016, I happened to open a spam e-mail from the National Institute for STEM Education (NISE) and registered for the organization's certification program. I knew little about NISE other than its association with STEMscopes, which had created a 41-part process to achieve the National Certificate for STEM Teaching (NCST). After digging in to the certification, I soon realized I had found the golden ticket for bringing our learning community together as STEM educators. I dug into each indicator and reflected on my classroom practices using this new framework while creating a portfolio of submissions on myriad topics. I found I was implementing some practices well, others could be done better, and there were definitely a few I had never even considered. Two weeks later, I completed the program.

After finishing the program, I had the formula to begin work in developing our CHS common understanding of STEM education, as well as the knowledge to provide professional development for my colleagues. Unbeknownst to me, I had just created multiple part-time jobs in addition to teaching my regular course load. Most of my fellow teachers and administrators loved the framework I had brought back to our learning community, and we have spent four years with it at the heart of our professional development. I led the initial delivery of these STEM education work sessions, but as others completed the NCST program, I shared the workload with capable coworkers who carry the torch to build our STEM learning environments. To date, nearly 50% of our educators have completed or are in process of completing their certifications, and I am fortunate to have coached all of them.

This process helped us pursue a National Certificate of STEM Excellence (NCSE) for our campus through NISE. We focused on three main goals as we moved toward our STEM vision. While facilitating this work, I was joined by colleagues and community members in creating our STEM vision and pathway to achieving it. After earning our NCSE in 2018, we are now at the end of our initial three-year plan and in the process of evaluating our progress and next steps. Having had the chance to work with an elementary school in Colorado and a charter school in Ohio, I can reiterate my earlier statement that no two STEM schools are the same.

While our program is far from complete and is always evolving, it has been built on the foundation of culture and leadership that makes the way we do things truly unique to CHS. Depending on when you come to our campus, you might see any combination of science and math teachers working with CTE teachers to provide instruction for students. These educators have been inspired to move away from the traditional modes of instruction. You can hear teachers saying, "What if we ..." when speaking to colleagues instead of remaining isolated. You would hear students asking, "Can we do this instead to demonstrate our learning?" instead of simply showing growth in learning by being good at memorizing and regurgitating information on a test.

So what has allowed our culture at CHS to evolve into having a STEM focus? From the humble beginning of this initiative, our culture has been about servant-centered leadership—the idea that the leader is a servant intentionally seeking to serve others first. Treating students as clients is no different an approach than building a computer network for a company or a house for a family. When you put your clients' interests first and approach every detail with a clear vision, the plans for how to proceed also become clearer. What follows are insights into the personal factors I believe allow someone to be a servant leader. While the five ideas I will explore are not by any means meant to be all-encompassing, they are concepts that have held up whether I was designing telecommunication circuits or working to build a STEM program in a larger public school.

Family, Loyalty, and Hard Work

I did not know this at the time, but I grew up extremely blessed. While we did not have much extra money and my friends were always out doing seemingly more exciting things than I was, it was far from a difficult childhood. There were always ample opportunities to spend long days working at my grandparents' farm, stepping up to help Mom out while Dad worked as a police officer, or pursuing any other opportunities to assist someone else. I now realize I was immersed in the process of learning one of the most important lessons of life: We cannot always dictate our situations when forced to do what is right, but we can always choose to do what is right based on our moral compass.

In retrospect, getting up early and heading for a long day of work at the farm served many purposes beyond the obvious, for example. This responsibility showed me that loyalty and family are some of the most important things we have in life. I learned that doing the right thing often means sacrificing the fun and easy options, but it also builds loyalty and purpose in ways not otherwise possible. As for leadership specifically, I internalized that when a leader is so loyal as to sacrifice his or her own interests for others, he or she will feel a sense of meaning and desire to work toward a solution to the problem at hand.

Creating a learning environment based on shared values, such as those within a family, as well as loyalty and hard work is intentional work that must be cultivated and cared for over time. The physical space of a classroom, the routines and procedures, consistent expectations, responses to failure, and ways to promote intellectual risk-taking are just some areas where a teacher can begin nurturing ideas associated with shared values, loyalty, and hard work. Don't be afraid to get into the work with your students and model what being a caring, loyal member of a hard-working team looks like.

Our STEM learning community is intentional about building the shared values of family and loyalty through our inclusion of students into the decision-making process. We constantly look for opportunities to engage students in meaningful learning situations. For example, for years our Advanced Placement Environmental Science teacher and her students envisioned building an outdoor classroom. There were hundreds of designs and redesigns produced and thought had been given to many, if not all, aspects of the structure. Students raised money for building this learning environment. It was everything you would expect of a grassroots effort—minus the final outcome. In 2018, I was approached about some money available for physical improvements at our school site, with the constraints that they needed to be designed by the spring semester and built during the 10 weeks of summer. These were pretty crazy timelines and difficult to meet, but the money was available, so we were off and running.

I gathered a group of older students and set out to work with them on implementing the two projects we decided to pursue, the outdoor classroom and a makerspace addition to our STEM building. At first, this was a labor-intensive process, as the students had no idea where to even begin. But after a month of working with the architect, engineers, tradespeople, and school district, the students slowly began to take over. My belief in the students led them to discover their own abilities. They would say things like "Don't worry, we've got this" and bring in people with skill sets or knowledge as needed to help them move forward. I slowly became a "proud father" watching these young people grow into thriving adults with the abilities to accomplish great things.

We adhered to the constraints of the project and opened both our STEM makerspace addition and the outdoor classroom in the fall of 2019. I was fortunate to be hired by the general contractor as a foreman and oversaw the building of these two venues. We succeeded in such undertakings in such a short time because we established shared values—including loyalty and hard work—in our STEM culture. I had complete faith that these learners would make the projects happen because I knew their loyalty to our vision and their willingness to do what needed to be done to realize the outcomes. Through the sense of loyalty and hard work, the educational setting can foster students' abilities to achieve seemingly unattainable solutions to problems.

Personal Meaning

As a graduate student, I read Dr. Viktor Frankl's book *Man's Search for Meaning* (Frankl 1962). I want to be clear that I would never compare anything I have been through with Frankl's World War II experience in a concentration camp. His work, however, certainly provides perspective on what is important in life through the lens of extreme circumstances. Frankl's sense of meaning is expressed in his summary that if you have a *why*, you can live with any *how*. This text has stuck with me over the years, but entering the field of education has allowed my focus to shift on the central concept of one's meaning in life. With time seeming to go faster, the concept of personal meaning weighs more on my daily life. When I was younger, I could not appreciate the fragility of life and the importance of personal meaning. Nonetheless, meaning has quietly been the foundation of everything I have done and has informed how I have approached others in the actualization of personal meaning in given circumstances. My time in the education space has been so eye-opening into the effect that meaning has on individuals and how they move through their journey.

Let's be honest: Education is far from a "get rich quick" profession. So why do we do it? We do it because most of us love to provide meaning to learners. In most of my educational career, teachers tried to provide meaning by standing in front of the classroom and talking. While I did take some things from these interactions, the teachers who allowed me to demonstrate growth in learning on my own terms and who brought their personal meaning to their teaching created the experiences that I can recall to this day.

Engaging students in authentic learning is the best way for students to own their learning and for teachers to differentiate between unique learners. Delivering authentic learning can be an emotionally draining endeavor because 25 students can be at 25 different places along their learning journey, and you are the GPS to help each one progress along their unique route. Delivering authentic learning that students take with them past a summative assessment is difficult, but our students deserve this experience and should expect it from each of us as teachers and leaders.

We have been fortunate to send our STEM graduates to universities such as Harvard and Johns Hopkins and into industry and trades. While one might seem a better outcome than the other, what truly matters are situation, school, and positive outcome for the student. STEM education should focus on the idea that one outcome is not better than another in most cases, but these outcomes can be equally correct for individual students' pursuits.

Controlling What You Can Control and Embracing Change

One of my bad traits is that for a long time I wanted everyone to like me and what I was doing. It turns out that I don't really want everyone to like me. What I really need is for people to possibly disagree with me while still respecting my hard work and moral standards. I admire my wife for many reasons, as she is someone I view as a strong leader who is comfortable in her own unique style. We do not agree on everything, but we respect each other for who we are and acknowledge that we can learn something from each other—if we listen.

"Control what you can control" is a phrase that is repeated in my mind, and the voice sounds eerily similar to my wife's. It hasn't always been there; without this voice, I have spent a considerable amount of time worrying about things I couldn't control. Over the past 20 years, the phrase has slowly become a staple of my everyday approach to life. To say I am consistently implementing this concept is inaccurate, but I am certainly better at controlling what I can control today than I was yesterday, and I will strive to do a little better tomorrow.

Understanding what is in your control and what is peripheral serves many purposes in leadership and for your personal sanity. We only have so much energy to expend in a day. What we choose to spend it on becomes a reflection of who we are. If we choose to spend time worrying about things we cannot control, we waste time and energy that can be spent on more meaningful endeavors. When others understand our sincere concentration on spending our time and energy on things we can meaningfully impact, they are inspired to do the same. Human nature will not allow most of us to ever perfect this concept, but as Vince Lombardi stated in his first team meeting as the coach of the Green Bay Packers, "Perfection is not attainable, but if we chase perfection, we can catch excellence!" (Family of Vince Lombardi n.d.).

When we talk about controlling what we can control, it is inherent that we must embrace change. We cannot control change, but we can control how we adapt and incorporate a changing environment. Having played and coached football for 20-plus years, I find the sport to be the epitome of this concept, as there are so many variables involved when orchestrating 11 people to solve a problem another 11 people are creating—then doing this 60 or 70 times per game. So how do you handle situations when everything you have prepared for disappears? How do you handle it when your teammate is suddenly injured and replaced by someone less knowledgeable? You have two choices: Quit, or embrace change to control what you can.

In our STEM environment, my goal is that every student will work through a personally authentic problem in a way that makes sense, with me serving as the servant-centered leader. If I have 25 students, I want 25 different things going on that I can facilitate to grow learners. This processes causes authentic learning environments to look, sound, and feel different than what we still envision for school. But the underlying theme is that learners are taught to control what they can, in addition to the problem-solving process and the embracing of change. If students remain true to this process, the learning outcomes take care of themselves. The STEM teacher takes a back seat to the learning and serves as the conductor refocusing learners on what they can control in given circumstances. In a paradoxical turn of events, the educator is in control by not being *outwardly* in control and instead allowing students to experience learning in a way that makes sense to them within the scaffolds and structures provided.

Put the Greater Good First

Whether you are a project manager, contractor, coach, or teacher, anytime you strive to bring a group together for the completion of a desired outcome, it cannot be singularly focused on the good of a single individual. I have held several positions, and each had its own unique set of circumstances. Variables among them have included amounts of money, the personal safety and physical growth of young people, and intellectual risk-taking and response to failure. The constant is that success must be approached with a *we* versus an *I* perspective.

Upon entering the coaching and teaching professions, I made the mistake of doing both the same way I had experienced being coached and taught. A friend who is a successful high school and college coach and educator guided me out of this way of thinking with his mentorship. It turns out that common ground exists between coaching and teaching in that it really doesn't matter what I know unless my students can also have that knowledge.

The idea of *we* in the STEM classroom comes to life through peer mentorship. By building and intentionally speaking to our inclusive STEM culture, we have created an environment where students not only are willing to help one another without regard to any labels but also take an immense amount of ownership in their STEM world.

The makerspace I mentioned earlier is an excellent example of the idea of putting the greater good first. Through grant writing and private local contributions, we have amassed an arsenal of technology and the ability to create strong facilities and programs. The problem students identified concerned access. If they had success in finding solutions to problems thanks to the equipment they could use, shouldn't others have the same access? From this discussion emerged the idea for a makerspace addition. While not huge, the 800-plus square feet that we added onto our STEM building offers a dedicated place where all equipment is accessible to students without interrupting learning in the classroom. We have had students from every content area visit the makerspace in order to demonstrate their learning in a way that is authentic to them, with STEM students assisting them. The makerspace has truly become an area of commonality.

Anytime you begin bringing others into the equation, you will have unforeseen challenges. Establishing clarity of vision for the whole group allows plans to be made based to achieve that vision. When plans are made in this way, the purpose and meaning are supported and the leader bears responsibility for making some decisions.

Appreciate Personal Experiences

People buy into a vision when they feel important and integral to the outcome. Personal experiences not only shape who individuals are, but when individuals are respected and intentionally brought into the conversation, everyone involved gains a helpful understanding of the situation. This fact was never entirely clear to me until I entered the classroom as a teacher. My great friend from college had done the same thing a few years earlier and talked about how much he loved the profession. If he could do it, why couldn't I? I had no idea how difficult some students' home lives are, no clue all of the things today's youth deal with, and no understanding of how emotionally draining it is to balance all of these variables with educating young people on a daily basis. Of all of the careers I have been blessed to experience, meeting the needs of and being a consistent presence for our students is the most difficult and rewarding.

The most effective way I found to build the relationships that are so vital to supporting students' needs in the learning environment is showing I care by listening to their personal experiences. I have been able to meet students where they are through an understanding of where they have been. This empathetic approach has taught me things I could not have learned any other way. There is so much that I have never been exposed to that these kids have that I would be a fool not to listen and grow my understanding by looking at the world through their eyes. I believe I would have achieved more academically as a student if the focus of my education had been on where I was and not where the curriculum pacing guide dictated I needed to be.

The STEM educational model we have developed at our school would not be where it is now without this respect for individual experiences and interests and the pursuit of those as a requirement of the curriculum. It is difficult at times to translate the learning that occurs in this environment to standardized testing, but there is no difficulty in detailing the learning that occurs to the world for which we are preparing our learners. When we consider that our youth need to be prepared for jobs we cannot even imagine and that they instantly have access to all the information they could ever need, what is really our job as educators? Our most important job is to nurture students' abilities to identify and apply information so they can navigate any problem-solving process and create solutions.

The STEM model at CHS has evolved to include capstone projects, which demonstrate growth in learning over a long period of time that would not have been possible with a student learning solely in a classroom. This loose definition allows everyone from our highest achievers to those who may have made poor choices and paid steep prices to share their learning journey with an audience. For those who need it, we can use this project as a partial graduation requirement; those who have worked hard can earn a Colorado Department of Education STEM-endorsed diploma. I have had the privilege of listening to students whose stories have brought tears to my eyes and who, without the opportunity for capstone work, would have been cast as failures. I have taken students to present before educators who hold the highest positions in Colorado's education system so they can speak on behalf of the benefits of our model of STEM education for their personal growth in learning. Quite simply, capstone work has changed our learners' expectations of their high school experience, and this work is predicated on the value placed on students' personal experiences.

This environment of intellectual risk-taking and positive response to failure develops skills vitally needed in our world today. The learning community should value and allow for personal experiences to be applied beyond the curriculum. Giving learners differentiated opportunities to apply their growth in learning in a safe and inclusive environment allows them to practice skills that will benefit them on their journey and provides the relevance necessary for authentic, lifelong learning.

Conclusion

The experience of being in public education, complete with hopes and dreams of making a difference, has reinforced in me the concept of servant-centered leadership. You cannot succeed in this profession without giving yourself to others. After years of schooling and personal experience, we are all able to present curriculum to learners and expect that they soak it in as if they were sponges. If that is where our work ended, we probably wouldn't feel so energy deprived at the end of each day. No longer are we the sage on the stage throwing information at eager learners who desperately wait for each word to come from our mouths so they can take the wisdom as their own. In the time it took you to read the last sentence, today's students are onto something else. I am not saying that we send up the white flag and surrender to the circumstances of our current reality, but instead that leadership in the learning environment must recognize and adapt to this reality.

By creating an environment where students feel they are a loyal part of a family, where they feel meaning in spending time and effort, where they not only expect but also embrace change and learn to control what they can control, where they practice seeing the *we* before the *I*, and where they respect and value one another's personal experiences, we empower students to move forward on their personal journeys. But all of this doesn't just happen magically. This beautifully chaotic and messy learning environment is the result of intentional leadership and, as such, is constantly morphing to fit the needs of its clients. Just as every family has unique dynamics, so, too, does every STEM learning environment, which is why so many have a difficult time wrapping their minds around what STEM education should look like. Here is what I can offer you: Being intentional to the five points I have discussed in this chapter will not give you a STEM learning environment that perfectly replicates the one we created at CHS. What you will create, however, is a STEM learning environment that perfectly looks just as it should for your learning community and all of its stakeholders.

RODGER'S INSIGHTS AND INTERPRETATIONS

Phil's profile demonstrates a leader's role as advocate, collaborator, modeler, and provider of resources. These roles easily and clearly emerged because of Phil's insightful theme of servant-centered leadership—the idea that a leader intentionally seeks to serve others first. The perspective, by its very nature, suggests a person-centered relationship. This view is confirmed by several of the five ideas he presented, such as providing a *why* and *how*—meaning that you should place the greater good first, appreciate others' personal experiences and abilities, and apply skills to authentic problems to ensure relevance and meaning.

CHAPTER

10.

My Path to Education Leadership

MAYA M. GARCIA

Maya M. Garcia was, until January 2023, a supervisor and science content specialist at the Colorado Department of Education in Denver, Colorado. Ms. Garcia is now Chief Program Officer for Beyond100K.

NOTE FROM RODGER

Maya Garcia leads the implementation efforts for Colorado's Academic Standards for Science, which was her initial responsibility when she joined the Colorado Department of Education in 2019. She also has provided leadership in responding to COVID-19, addressing climate change, and enhancing equity both in Colorado and nationally through her contributions to the Council of State Science Supervisors.

Maya began her career as a middle school science teacher and science department chair in the District of Columbia, where she taught for more than eight years. In 2013, Maya joined the District of Columbia's Office of the State Superintendent of Education. She led the District's implementation efforts for the *Next Generation Science Standards* and the *Common Core State Standards for Mathematics* and developed the District's plan for advancing preK–12 STEM education.

Maya earned a bachelor's degree in neuroscience and behavior from Mount Holyoke College and a master's degree in science teaching from American University.

Maya is already a leader in the science education community, as demonstrated in her profile.

My Early Years

As a leader in education, I learned early that I needed to understand how my own path manifested and to be curious about the experiences that have influenced the people and communities with whom I work. As I reflect on my journey, I have clear evidence of how my relationships and interactions have shaped and will continue to shape who I am as a leader.

My childhood in Los Angeles was liberating. I played with friends, walking miles before spending afternoons reading in libraries or bookstores, discovering parks and botanic gardens, and helping my dad with our booth at the Pasadena Community College Swap Meet. My parents, both Los Angeles Community College (LACC) graduates, were actively engaged in my school and our community outside of their full-time jobs. Afterschool activities included meetings of the Parent-Teacher Association (PTA), city council, Los Angeles Unified School District board, and even unions. You could say my siblings and I were "professional" meeting attendees and campaigners. At these meetings, I watched my parents work alongside other parents, teachers, and school administrators to ensure students attending our schools had access to the best the district could offer. I watched as my father, a gardener and pest control supervisor for the school district, grilled a team of lawyers from the city about school funding and education rights for English learners. I watched my mother navigate the nuances of specialized education to learn about better supports for my brother, as well as her work to develop more assistance for African American teachers and students. My parents ensured that their three children were immersed in conversations about privilege, educational equity, and race from the moment we could walk. We learned early to question policies related to issues such as tracking and to confront bias and injustice in education.

We also learned the importance of reaching out in the community to ask for help and to be generous with help in return. To dismantle and challenge the systems meant to marginalize our community, we needed to organize with others and learn to share resources. We learned to value literacy, the natural world, civics, language, and community.

My childhood years influenced my approach to classroom teaching, leadership, equity, and justice. As a woman of color in STEM, I recognize that these early experiences influenced my priorities and core values as an educator, as well as my leadership at local and national levels. These experiences formed my roots. I learned lessons about advocacy, community organizing, accessibility and language access, and the importance of being in situations where I had a voice at the table.

Becoming an Educator

During my K–12 education, I was placed in honors courses and had amazing science teachers who pushed me to pursue a career in STEM. From a young age, I was encouraged to tinker and be curious, and I had a healthy relationship with the natural world. I wanted to be a veterinarian or a doctor from a young age. In middle school, I was placed in advanced math courses and discovered a talent for and interest in science, with the benefit of having stellar female science educators who served as strong role models for a young woman of color. During my junior year in high school, my chemistry teacher, Mr. Neth, pushed me to apply to the NASA SHARP PLUS program, and that summer, I traveled on an airplane for the first time and joined 19 other students from across the country at the University of New Mexico to participate. We all interned in departments or at national laboratories and learned to code. The experience was transformational. I connected with students who had a passion for science and engineering, and this program exposed me to a different way of engaging in science, one that was not tied to "cookbook" labs or a set curriculum, but instead encouraged me to think beyond, wonder, and ask deeper questions. The experience was empowering on so many levels, as I realized that I did not have to wait until I had a degree to "do science." I also remember wondering why my high school classes were not structured this way.

As I transitioned into my undergraduate work at Mount Holyoke College, I struggled with my major in biochemistry; even though I persisted, I did not really have a passion for the subject. I ended up taking a year off, much to the shock and dismay of my family, and worked at Ramapo Anchorage Camp, an organization that serves children, youth, and families with special needs. Camp Ramapo served more than 500 children with special needs, ages 6–16, providing an opportunity for children with a wide range of needs—from nonverbal elementary-age children to teens who needed help building conflict resolution skills—to experience success. I was invited to stay on to work as part of the year-round staff. I collaborated with the rest of the team to develop and lead large- and small-group programs and facilitate ropes courses for schools, community-based organizations, and families, mainly from New York City. I learned about developing inclusive and empowering programming for youth, the value of experiential learning opportunities, and how to plan programs that support trust-building and effective communication. I also learned that I had a passion and desire to work with children and families in education. I returned to Mount Holyoke College, completed a degree in neuroscience and behavior, and applied to the D.C. Teaching Fellows alternative certification program. I was accepted into the program and placed in a secondary school science teaching cohort.

These formative experiences during high school and college informed my work and values as a science educator. When leading professional development, I sometimes ask participants to consider their own relationship or connections to science. It helps to be able to reflect on your journey and center thinking on the types of learning experiences that stay with us, both positive and negative. As a classroom teacher, I developed lessons that were rooted in real-world experiences in ways that would empower my middle school students. My work was influenced by my learning in outdoor education, my background in neuroscience and behavior, and my experiences with differently abled students and their families. I created learning experiences in which students were "doing science" in inclusive and empowering ways. My experiences helped me recognize the connections among instructional design, learning environments, and student identity, and I carried these lessons into my role as a leader.

An Emerging Leader in K–12 Education

My journey as a leader in K–12 education began with my work in the District of Columbia. As a novice science teacher, I was invited to join the state science standards review committee so I could offer a new teacher's perspective. As the youngest member of the team revising science standards, I met veteran teachers, members of the D.C. State Board of Education, and university partners. This experience anchored my future work in science education. I noted how each committee member contributed, and through intense discourse, we developed a common vision and values. I recall being reticent to share my input as the novice in the room and would often defer to my veteran colleagues in discussions. The committee chair approached me after our third meeting and shared that each voice at the table brought a critical perspective to the conversation and the diversity of opinions would strengthen our overall approach. He also noted that by not sharing my opinion, I was hindering the process. I became a vocal participant and learned that my questions and concerns were shared by many in the group. I was honored to work with such fabulous teachers on that committee, many of whom served as mentors.

After that experience, I was asked to lead professional development for the new standards and support D.C. Public Schools with curriculum development and an aligned scope and sequence. Over the next few years, I went on to chair our school's science department, serve on school leadership teams, and support curriculum and professional development for science educators as a teacher leader. I also had the opportunity to work as a master teacher for the Center for Inspired Teaching and at the Carnegie Academy for Science Education at the Carnegie Institution for Science, in addition to working as an adjunct faculty member for American University's School of Education. These opportunities enabled me to learn more about adult learning theory, the design of teacher leadership pathways, and how district and school level policies are shaped.

In 2010, I was selected by the U.S. Department of State as a Fulbright Distinguished Teacher. I traveled to South Africa and conducted action research for six months. During that time at the University of KwaZulu-Natal, I observed how partnerships between cultural institutions, science organizations, and other institutions were leveraged to expand access to teacher training and hands-on resources for under-resourced communities.

When I returned to Washington, D.C., members of the committee on standards were deeply engaged in conversations about the *Common Core State Standards* and the *Next Generation Science Standards* (*NGSS*). As one of the leaders of the District of Columbia Science Teachers Association, I worked with my peers in the initial review process. At this time, I joined the Office of the State Superintendent of Education to lead *NGSS* implementation efforts.

A Teacher Leader in STEM Education

My role as a teacher leader influenced how I engaged with classroom teachers as a policy leader at the state level. I knew it was crucial to engage classroom teachers as leaders of our adoption and implementation process for the new state science standards. In 2013, I launched the first District of Columbia Science Teacher Leader Cadre, a team of educators who helped design the vision and strategy for our state's implementation plan. Educators served as ambassadors, professional development providers, and sources of critical feedback about our communication, assessment, accountability, and parent and family engagement strategies. During my tenure as the director of STEM for the District of Columbia, we launched teacher leader cohorts. We set collective targets and goals as part of this process, which allowed us to establish strong feedback mechanisms to help us revisit and adjust our efforts. This process of collective visioning and goal setting with diverse stakeholders set the tone for the launch of our state STEM plan implementation efforts. We launched annual STEM Summits with our partner, the Carnegie Academy for Science Education. At each Summit, we set collective goals for advancing STEM education in the city and highlighted work being done by cross-sector partners. In 2016, we launched the STEM Ambassadors program as part of our STEM Learning Ecosystems initiative to engage parents, community organizations, and students in the implementation of our state's STEM plan. Equitable community engagement and involvement are key to advancing a vision that results in inclusion for all students, a core value that was cultivated in me from a young age. I had the privilege of working alongside community leaders and educators who helped elevate, support, and sustain that work. In this role, I learned about navigating the tension between innovation and implementation.

As a leader in the policy space, there is an art to implementation efforts that involves weaving together multiple conditions, understanding the variables that influence outcomes (e.g., implementation fatigue, fiscal and budgetary constraints, leadership priorities, public sentiment, policy changes, etc.), and learning how to navigate these shifts and make course corrections as needed. Having so many factors in implementation shows why it is so important to develop community buy-in, ensure diversity of voices at the table, and set collective goals and establish feedback mechanisms.

As a teacher coming into the policy space, my learning curve was high. I learned quickly that the skills I leveraged in the classroom would serve me well. It was intuitive to engage educators in the planning process. Professional learning experiences felt like lesson planning for students; building consensus and developing strategic goals were iterations of unit planning with a team. All the skills I leveraged as a classroom teacher and the relationships with colleagues and organizations that I had honed over the years helped me as a leader and continue to be the foundation for my work at the local and national levels.

Leadership at the State Level

In 2018, I joined the board of the Council of State Science Supervisors, a national organization for state science leaders. During my three-year term, we worked to implement a position statement on equity. In this capacity, I served as a co-principal investigator and collaborated with the University of Washington and the University of Colorado Boulder on the Advancing Coherent and Equitable Systems of Science Education project, which supports state leaders in designing more equitable implementation approaches for science education. I have also been selected to serve on the Board on Science Education for the National Research Council, and in 2019, I became the science content specialist at the Colorado Department of Education, where I work with the community to support implementation efforts for the Colorado Academic Standards for Science. I truly love what I do and think of myself as a reflective practitioner and a constant learner. As I reflect on my professional experiences and consider what informs my work as a leader, I see connections to my childhood in Los Angeles and the influence of my family, former teachers, colleagues, and community, as well as the learning experiences I have had as an educator and a policy maker. Ongoing learning is key to evolving as a leader.

Conclusion and Next Steps

I have learned valuable lessons about the importance of mentorship, ongoing learning, community conversation, and strategic planning. In Washington, D.C., I was promoted into a supervisory role to support the District's larger STEM education efforts, and I was selected to participate in an executive leadership course, where I was introduced to organizational leadership, transition management, strategic planning, and strategic approaches to talent management. The course was a brief introduction, and I quickly recognized that my training and education were lacking and I had a desire to learn more. I started curating a list of authors, subscribed to *Harvard Business Review*, invested in executive coaching, and tried to advance my understanding of systems leadership. At this time, both my supervisor and my mentor counseled me to pursue an advanced degree. Each shared their own experiences as women of color in leadership, citing their doctoral work as transformational influences in their leadership journey and the value of learning with a cohort of peers. After conversations with my mentors, friends, and family, I recognized that it was time to pursue an advanced degree. With its focus on equity, the doctoral program in education at the University of Colorado Denver was ideal. I selected a focus on executive leadership to deepen my understanding of the organizational levers that are critical to the development of equitable learning systems. As I moved into my second year of graduate study in 2019 and 2020, I was challenged to analyze issues from multiple perspectives and to introduce new research and theory to my practice, both of which help me refine my leadership approach. I am excited to continue my journey as an educator and a leader and look forward to adding more to my "tool kit" in service of more equitable schools and systems of education.

RODGER'S INSIGHTS AND INTERPRETATIONS

In Maya's descriptions of experiences and opportunities that contributed to her growth as a leader, one can identify core values of advocacy and equity. Although her experiences were varied, Maya adapted the knowledge, values, and skills she learned as a classroom teacher. For example, she noted instructional design, the learning environment, and student interest and inclusion as key points that also apply to professional development.

Although Maya is modest in her reflections on leadership, one can readily identify several insights that aspiring teacher leaders would benefit from considering:

- Leadership begins with a clear vision and strategic plan.
- If you want to be a voice at the table, speak up.
- Have the community participate in the process of change.
- Be aware of and listen to mentors.
- Do not lose sight of equity as a goal of education.

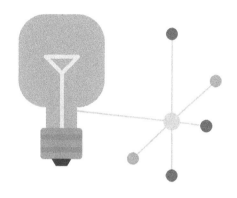

PART III

Some Leaders Support Science Teachers by Providing Professional Learning

11

Challenges and Leadership in Science Teacher Education

HERBERT K. BRUNKHORST

Herbert K. Brunkhorst is a science education professor (emeritus) at California State University, San Bernardino, in San Bernardino, California.

NOTE FROM RODGER

Teacher education for undergraduates sets the foundation for leadership in classrooms and the larger education community. Herbert Brunkhorst spent his career as a leader in various capacities related to the professional learning of future teachers and those currently teaching science. Herb's profile includes examples of national policies and the politics of college and university programs.

Herb has an undergraduate degree in biology and chemistry from Coe College in Cedar Rapids, Iowa. He continued his education at the University of Iowa, graduating with master's and doctoral degrees. He completed additional educational work at Harvard University and Stanford University. Herb has received the Robert H. Carleton Award and the Distinguished Service to Science Education Award from the National Science Teaching Association.

Early Years

My older sister and I are the children of immigrant parents who came to the United States from Germany in the late 1920s. My mother only went to school through eighth grade, and my father was trained as a machinist. They believed in getting an education, learning, working hard, and doing your best. They valued humility, honesty, patience, and perseverance. I provide this information as context for what I will share in my profile.

I learned responsibility and leadership early, serving as a member of the Jefferson Elementary School Safety Patrol in fifth and sixth grades and having a sizable paper route for which I delivered papers every afternoon and collected payments. In junior high and high school, my leadership opportunities centered on music, athletics, and scouting. I joined Boy Scouts at the age of 11 and attained the rank of Eagle Scout at 13, making me one of the youngest ever to attain that rank in New Jersey. My role as patrol leader and senior patrol leader taught me how to cooperate, delegate responsibility, and lead individuals with diverse personalities. In high school, I served in student government, in addition to being president of the Booster Club and co-captain of the soccer team during my senior year. I also participated in wrestling, track, marching band, and concert choir and was the photography editor for the yearbook my senior year. Each opportunity provided leadership experience because I was part of a group producing a product, whether that was a yearbook, musical performance, or sports victory. Two faculty specifically commented on my leadership abilities. Madeline Terra, the yearbook faculty adviser, wrote in my yearbook, "Herb, we are deeply grateful for the fine job you did on the yearbook. We will sorely miss leadership such as yours." The choir director, Don Angelica, wrote, "Thank you sincerely for all of your help all year long. Your leadership abilities were invaluable to me, and you helped make this year's choir a huge success." Mr. Angelica became a mentor and almost convinced me to major in music in college.

There is one more personal situation I need to share from my high school education because it relates to a renewed respect for my father. I was working on my trigonometry homework at the dining room table when my father leaned over my shoulder. He infrequently, if ever, showed an interest in my homework. But this time he said, "Oh, you do that stuff too!" I had never made the connection that as a machinist my father worked with trigonometric functions every day. This moment gave me a better understanding of and respect for my father and his humility, quiet ways, and education.

I attended Coe College, a small liberal arts college in the Midwest, and was exposed to a new part of the country. I had three mentors at Coe who became colleagues and friends after graduation. My first exposure to real inquiry teaching was from Dr. Robert Drexler. Whenever you asked Dr. Drexler a question, he would respond with a question; before you knew it, he had you coming up with the answer to your original question. He was a mentor and adviser who encouraged me to take a variety of courses outside science. His philosophy was "You want a variety of balls of yarn in your basket because you never know which direction you might need to go." It provided the kind of liberal arts education I value today.

In the summer between my junior and senior years, I was invited to be a teaching assistant for my physiology professor, Dr. Kenneth Cook, who was teaching introduction to biology during the summer session. He asked if I would take responsibility for the laboratory and later asked if I would be interested in teaching a few of the lecture classes. I took advantage of the opportunity, and a whole new world opened up to me and led me to the teaching career I have enjoyed.

I became a teaching assistant during my senior year with Dr. Cook in physiology and Dr. Ruth Siemer in bacteriology. Both became mentors and colleagues during that time and longtime friends after I graduated from Coe. I was asked to teach the entire summer session introduction to biology class and laboratory following graduation. My early experiences taught me to how to be a better listener, have greater empathy, and, in several cases, become a mentor.

That summer, I applied to graduate school at the University of Iowa, where I became the teaching assistant to Dr. Robert E. Yager at the University Lab School. This opportunity resulted in a collegial and personal friendship that lasted 53 years, until his death in 2019. He was the most influential mentor in my professional life.

Early Lessons in Leadership

I began my career by serving as the director of the Center for Science and Mathematics Education at Weber State University in Ogden, Utah, where I was responsible for the annual regional science fair. The center's associate director and I decided to implement a more equitable evaluation system for the science projects to ensure a more diverse group of students would succeed. Little did I realize that we were upsetting the status quo, and the next year, I was relieved of my duties as director of the science fair! This experience provided a lesson about politics in leadership.

After joining the faculty at California State University, San Bernardino (CSUSB), I became the director of the Institute for Science Education. The institute had university-wide recognition, which allowed it to include faculty from other colleges throughout the university. The hope was that funding would come through the institute and be distributed to the colleges and departments involved, with the expectation that this process would make grants more competitive because of the potential for broader expertise across applicants. As it turned out, the institute ended up being another source of competition within the university. The reward structure on our campus (and on most university campuses at the time) mitigated against this type of collaboration. Grants were coveted by individual colleges and departments within the university. This lack of a collaborative spirit eventually doomed the institute and offered me another lesson about the politics of leadership.

From 1988 to 2005, the Inland Area Science Project (IASP) was part of several regional professional development sites for science throughout California, with funding provided by the University of California's Office of the President. IASP was a shared collaborative between the San Bernardino County Superintendent of Schools; the Riverside County Office of Education; the University of California, Riverside (UCR); and CSUSB. The fiscal agent for the project was housed in University Extension at UCR. The project offered two three-week sessions of professional development for K–12 teachers of science and included an academic-year component with teacher stipends. Professional development was provided in biological, Earth, and physical sciences along with the BSCS 5E Instructional Model. In the mid-1990s, a component was added to IASP to address equity in science. Several of the early participants in the project became part-time instructors, and several became science leaders in their school districts. The project led to a sense of unity among the area's science teachers, and we frequently held a barbecue pool party at the end of the summer program to encourage this camaraderie. The program ended when the team of directors attempted to have the program moved on the campus of UCR to the biology department, which caused quite a controversy. This was another experience that offered a lesson in leadership with regard to how money (and who on campus has it) frequently determines a program's success and stability.

In 1989, CSUSB became the preservice and professional science teacher development center for the National Science Teaching Association's Scope, Sequence, and Coordination project. Using funding available through the California Department of Education, CSUSB science education faculty established teams of triads (consisting of a secondary science teacher, science faculty member, and science education faculty member) on each of the 21 campuses in the California State University system at that time. Each triad received funding to carry out a science teacher preservice or professional development project on its campus under the leadership of the triad. Campuses submitted their plan before funds were distributed. A positive result of this project and its structure was that a generation of science teachers became leaders in the state and, in some cases, administrators in local school districts, including superintendents.

In the late 1990s, I was invited to serve as co-chair of the Committee on Science and Mathematics Teacher Preparation for the National Research Council. The committee comprised accomplished academic faculty from science, mathematics, and technology who had strong personal perspectives on education in these fields. By fostering mutual respect, maintaining a calm focused demeanor, and having lots of patience, I was able to coalesce the committee to produce a seminal report titled *Educating Teachers of Science, Mathematics and Technology: New Practices for the New Millennium* (NRC 2001). What I learned in this leadership experience enabled me to participate in and navigate two additional projects for the National Academy of Sciences that broadened my leadership skills with diverse groups of proven leaders in science, mathematics, and education.

While serving as president of the Association for the Education of Teachers of Science (AETS), I represented the organization on the Council of Scientific Society Presidents. Whenever I introduced the name of the association, eyes would glaze over by the time I reached the second preposition, and I would often hear "Who?" At the annual AETS meeting that year, I challenged the organization to think about a name change that would be more concise and inclusive. The membership and board took the suggestion under advisement and, after debate and a vote, instituted a name change to the Association for Science Teacher Education (ASTE). The organization now includes teacher educators, scientists, science coordinators and supervisors, and informal science educators who prepare and provide professional development for teachers of science.

My election to the National Science Teaching Association (NSTA) Board of Directors as director of preservice teacher education presented some new leadership challenges. The organization was considering leaving the National Council for Accreditation of Teacher Education (NCATE) because of large costs for little return; however, some states required NCATE accreditation. The board decided to walk a fine line by developing NSTA preservice standards while providing NCATE workshops for those science educators whose states required NCATE accreditation of teacher education programs.

In 1996, the College of Education at CSUSB was reorganized, and I became the first chair of the newly formed Department of Science, Mathematics, and Technology Education. The department included faculty from science, mathematics, technology, and career and technical education (CTE). I recommended establishing a master's degree in STEM (science, technology, engineering, and mathematics) education. This effort became the greatest leadership challenge I faced as department chair. Because we had science, mathematics, technology, and CTE faculty, I thought that if we combined our respective expertise and resources, we could offer a program in which all were involved. The missing component was an engineering program on campus, but I hoped that CTE faculty could fill the gap until we could hire an engineering educator. The CTE faculty could provide some reasonable engineering experiences and the rest of us could learn. Although we were successful in getting the program approved and started, it ended up not being sustainable. The lesson for leaders is that you need a critical mass of knowledgeable, confident, enthusiastic, and passionate individuals to sustain innovative initiatives.

I encountered another challenge as department chair when I was handed 10 theses in environmental education that had been rejected by the graduate dean on the basis that they did not represent master's-level scholarship. Our environmental educator retired due to this issue, and I had to help 10 students restructure their theses. I was reminded of my mentor Bob Yager's gentle admonition: "Focus on a sharply defined problem; you have the rest of your career to save the world!" I was able to mentor these students to develop doable research questions, get their proposals approved by the Institutional Review Board, carry out their research, and produce a final product that was acceptable to the graduate dean.

My final leadership challenge arose when I was co-chair of the California superintendent's STEM Task Force, which was charged with summarizing the status of K–12 STEM education in California. The task force consisted of 50 individuals, including K–12 STEM teachers, school administrators, university STEM faculty, business and industry representatives, informal educators, parent representatives, foundation representatives, and media representatives. Participants were assigned by areas of expertise, with some individuals assigned to related groups to provide cross-pollination of ideas. The entire process was—to use an old phrase—like herding cats. The task force met regularly over the course of two years, and in 2014, we issued the report *INNOVATE: A Blueprint for Science, Technology, Engineering, and Mathematics in California Public Education.*

Reflections on Leadership

My personal ideas about effective leadership include, but are not limited to, developing mutual respect, valuing all individuals, identifying people's strengths, and being an effective listener and communicator. Leaders must remain open to new ideas, innovations, and out-of-the-box thinking. Don't be afraid of failure and having to return to the drawing board. Never stop being a learner! Be realistic and fiscally judicious. Don't expect remuneration for everything you do, especially when assisting students. Have fun! Don't forget to work hard, but play as well. For the 16 years I was a department chair, our department was the envy of the university because we had a party at the beginning of the academic year, a memorable holiday party, and an end-of-the-year party. Those events helped develop a caring community among our faculty. Be generous in your praise and tempered in your criticism. Deal with problems while they are manageable and small. Listen to all viewpoints and exercise patience. Be willing to apologize and admit when you are wrong. Realize the limits of your own knowledge, and don't hesitate to ask for help or advice.

My advice to the next generation of science teachers would include to be passionate about your subject and helping students and to be a lifelong learner. Don't let obstacles get in the way. Remember the primary role of your administrator is to be a facilitator who can help you do your job better. Make your classroom an exciting place to be, where students are eager to learn. As a mentor of mine once said, "Be a guide on the side, not a sage on the stage." Aim high—there is nothing wrong with an elementary teacher having a PhD. Have self-respect and respect others. And put your students first!

RODGER'S INSIGHTS AND INTERPRETATIONS

My experiences with Herb's leadership confirm his integrity and the values he learned from his parents—humility, honesty, patience, and perseverance. At the conclusion of his profile, he offers guidance for leaders: Put outcomes for students first (e.g., create an engaging classroom to facilitate learning); pay attention to personal relationships (e.g., show mutual respect, value all individuals, listen first and respond second); remain open to new ideas (e.g., find innovations for standard thinking); develop a caring community among team members (e.g., work hard, play, be professional and social); deal with problems when they are small (e.g., seek help if required); and don't stop learning.

CHAPTER

12

Teacher Leadership

Personal Experiences, School Partnerships, and Professional Perspectives

TAMMY WU MORIARTY

Tammy Wu Moriarty is the associate director at the Center to Support Excellence in Teaching at Stanford University in Stanford, California.

NOTE FROM RODGER

Tammy Wu Moriarty is the associate director of the Center to Support Excellence in Teaching (CSET) at Stanford University. Tammy's work focuses on conducting professional development for classroom teachers, providing instructional coaching, and developing teacher leadership. Her experience includes being a secondary science teacher, district science resource teacher, school administrator, and educational consultant. She holds a bachelor of science degree in animal physiology and neuroscience from the University of California–San Diego, as well as a master's degree in educational leadership and a doctorate in leadership studies, both from the University of San Diego.

Early Years

I grew up in San Jose, California, the heart of the Silicon Valley back when there were still orchards lining the streets and Apple was a promising new business just down the street from my school. As a first-generation Chinese American, I benefitted from all of the privileges and opportunities that were offered to me, but I also experienced all of the prejudices and judgments that came along with being the child of immigrants from Hong Kong.

I did well in school, as my parents expected, and mostly enjoyed school. To this day, I can still name every elementary and middle school teacher I had. In elementary school, I was excited about learning new things and happiest when teachers took an interest in who I was as a person. In middle school, I went through the same awkward stages as everyone else and felt all of the insecurities that come along with that developmental stage. Those early years, however, were some of the most formative years in my life and probably started my interest in leadership. In seventh grade, I took an elective course called Leadership. Teachers recommended students to take the course based on their perceived potential to be a leader. I do not remember the content of this class as much as I remember the people in it. The teacher created a class community that made me feel like I belonged. I was in this class because someone thought I could be a leader, which gave me more confidence. I developed enough confidence that I took a risk the next year and ran for eighth-grade student government. I did not win and felt humiliated and like I had failed, which ended up being another important part of my early development as a leader.

In high school, I had to figure out who I was and where I belonged. For a brief time, I realized that I didn't really belong anywhere in particular. I did well in school, but I never considered myself the best at anything. Once again, I took a risk and ran for student government. I lost again and felt just as humiliated as the year before. Nevertheless, I found other ways to practice leadership. I joined clubs and took positions of leadership within these smaller settings. I volunteered my time both in and out of school, interacting with students from different grade levels and adults in various settings. I was involved with a church youth group and volunteered at the hospital where my mother was a nurse. I became really good at navigating new spaces and interacting with different groups of people. I still had a strong desire to run for student government *again* during my senior year (my third try in five years), and I finally won an election. These early risks and failures helped me develop resilience and offered lessons I would take into adulthood.

In college, I sought opportunities to get involved and be part of a community. I joined a sorority and took on leadership responsibilities in that setting, making decisions and learning how to navigate the social dynamics of more than 100 collegiate women. As I finished my undergraduate degree and moved into young adulthood, I became more involved with community service, volunteering with youth groups, children's hospitals, organizations that built homes, and a number of different projects. I participated in these activities because I enjoyed them. These choices would end up shaping my career.

Early Career

The experiences in my early years led me to become a science teacher. The various volunteer experiences made me realize that the kind of setting that would make me happiest was in education. I became a secondary science teacher and started my teaching career in middle school. Like most, I struggled during my first two years of teaching, trying to figure out how to handle all of the responsibilities that came with teaching energetic middle-schoolers. I had plenty of mentors help me through these struggles, and I leaned on them to help me improve what I was doing in my classroom. The support of these mentors helped me figure out how to be the teacher I wanted to be. I learned how to build a strong classroom community and loved my students. I began to take on more leadership responsibilities at the school, including leading science department work. I was noticed by individuals in the district and chosen as Middle School Teacher of the Year for the San Diego Unified School District. Although I felt truly honored, this spotlight was uncomfortable for me. This event did, however, open new doors for me.

The next year, I received a note in my mailbox with a highlighted job posting for a district science resource teacher. The note written on the posting said, "Tammy, I think this is something you should consider." It was not from my principal, but from our head administrative assistant in the front office, the person everyone knew kept the school running smoothly. I put the note away because I already loved what I was doing and didn't want to leave my school. The risk-taker in me, however, wouldn't let me forget about it. After some thought, I figured I had nothing to lose and that it wouldn't hurt to apply. I interviewed and, to my surprise, got the job. I had to sit with that job offer for a little while. I was sad to leave the classroom but thought this was a chance for me to grow and learn. I knew there would always be a place for me in the classroom.

I soon found myself surrounded by a group of science teacher leaders who collectively had the responsibility of furthering science teaching and learning for the entire school district. During this time, I had the amazing opportunity to meet the folks at BSCS (Biological Sciences Curriculum Study), as members of our science office became a part of a BSCS program called the National Academy for Curriculum Leadership. We were committed to professional learning over the course of several years as we learned from and collaborated with BSCS and other science education leaders from across the country to support high-quality science instruction in our districts. The years in the district office were pivotal in my learning. BSCS developed my understanding of the BSCS 5E Instructional Model, inquiry-based science, and how to create momentum around high-quality teaching and learning. I made connections that would end up being lifelong and life-changing, including becoming part of the K–12 Alliance, presenting at local and national conferences, and connecting with science educators from across the country. Little did I know how much this work would affect my future career.

Once again, my career took a slight turn as I went back to school to purse my master's degree in educational leadership and my administrative credential at the University of San Diego. I soon became an assistant principal and was back at the school again. Everything I learned during my time in the district office influenced how I approached my work at the school, including what it meant to support teachers in their work and how to navigate district systems. These experiences also opened my eyes to the complexities of school administration and what it takes to run a school. These years were challenging, but once again I was fully supported by amazing mentors who helped me approach the work in ways that felt manageable and sustainable. I was encouraged and had a strong desire to further my learning and go back to school to pursue my doctorate in leadership studies.

Later Years

As I was finishing my doctorate, my family and I moved back to the Bay Area and I returned to my childhood neighborhood. A few weeks after I officially defended my dissertation, I received a phone call from Janet Carlson, the former executive director of BSCS, who informed me that she had taken a position at Stanford University's Graduate School of Education. I applied for and was offered a position at the Center to Support Excellence in Teaching (CSET). I would once again be leading professional learning for science teachers. It was a job opportunity I could not turn down, as I would again get to do what I loved. I became a lead science educator and worked directly with teachers from across the country, leading professional learning sessions, providing feedback on classroom instruction via a video platform, and working alongside teachers in a coaching relationship. Being at Stanford's Graduate School of Education also provided continuous opportunities for me to grow and learn in a thriving intellectual community.

I soon had the opportunity to collaborate with others in CSET to develop a teacher leadership program for alumni. As I worked with a team to develop this program, much of the content and foundational ideas that I offered drew from what I learned in my doctoral program, as well as my personal experiences working at the school district office and as a school administrator. I incorporated aspects of organizational leadership, adult development, and group relations as key components of the work. My work soon spread to include developing learning opportunities for school leaders and teachers, which allowed for a systems approach to creating excellence in teaching and learning. Opportunities continued to present themselves within CSET, as teachers and schools worked toward programs aligned to the *Next Generation Science Standards* (*NGSS*; NGSS Lead States 2013). We expanded our leadership work by engaging with the Wipro Science Education Fellowship Program, which involved a more systemic, long-term relationship with local school districts.

Every aspect of my education and career experience has helped me develop into the person I am today, with a focus on the two professional areas that I care deeply about: science education and leadership. As the associate director of CSET, I continue to look for opportunities to further this work and provide the same kind of support and mentoring that I received throughout my career.

Developing Leadership Through Partnerships: An Example

What does *partnership* truly mean? More specifically, what does it mean to have meaningful partnerships with school districts that support the development of both teacher leaders and district leaders? This section summarizes CSET's partnership with the San Francisco Unified School District (SFUSD), a large urban school district serving a diverse student population, and how this partnership helped promote high-quality science teaching and learning in the classroom and build teacher leadership capacity in the district. I worked with Sharon Parker, a senior science educator at CSET, and Eric Lewis, a secondary science content specialist for SFUSD, who provided content for this section.

How the Partnership Started

In 2016, SFUSD needed to change its approach to science instruction prompted by the *NGSS* and district leaders' focus on finding core curriculum that would support *NGSS*. CSET was asked to help SFUSD's science leadership team learn about the BSCS 5E Instructional Model. Subsequently, CSET provided a series of professional learning experiences to help the science leadership team gain a better understanding of what the 5E Model would look like in the classroom.

Because of their basic understanding of the 5E Model, the SFUSD science department committed to finding instructional materials that had the following elements: the five components of the model (Engage, Explore, Explain, Elaborate, and Evaluate), problem-based learning, group work, and the use of local phenomena. SFUSD examined some curricula and found them lacking. SFUSD had a partnership with the Stanford Center for Assessment, Learning, and Equity (SCALE), which was developing a curriculum and noticing increased learning success with students. The SCALE curriculum was of interest to the district, but the developers only had materials for sixth grade. Based on its commitment to finding instructional materials with particular elements, SFUSD moved forward with SCALE and CSET to jointly develop seventh- and eighth-grade curriculum units that emphasized problem-based learning, local contexts, and the addition of the 5E Model.

Developing Leadership Capacity of District Science Leaders and Teacher Leaders

SFUSD's goal of creating a curriculum that would help reduce the opportunity gap for their students, especially historically underserved groups of students, was of utmost importance. In developing the curriculum, the team needed to ensure that students would be provided with engaging and relevant phenomena that were novel to *all* students.

The process of developing the curriculum through this partnership work allowed for thoughtful and intentional interaction between CSET and SFUSD. These interactions helped the SFUSD science leadership team understand the 5E Model in a deeper way; as a result, the model grew in importance and value for them. They soon found that the use of the 5E Model provided opportunities for students to create their own scientific understandings by building on their own prior knowledge and early experiences. Through a variety of experiences and explorations grounded in problem-based learning, students could refine and provide evidence for their ideas over time. Students who had felt disconnected from their learning because of having "unscientific" ideas or not having background knowledge were no longer at a disadvantage; instead, they were encouraged to share ideas and teachers were taught to acknowledge them.

This shift in teaching and learning brought on leadership challenges for the district science team because the change required teachers to interact with the curriculum and their students in different ways. No longer were teachers expected to disseminate large amounts of scientific facts and ideas for students to memorize. Instead, teachers were expected to listen to students' conversations and review students' work to identify what students were thinking. The work of teachers then became more about figuring out what experiences to emphasize to help students challenge initial or nonscientific ideas and foster more scientific ones. This shift was challenging for teachers, as they had to develop their own skills in listening to students, identifying students' current ideas about a concept, and figuring out the best next steps to support students' learning. The shift in classroom interactions was equally a challenge for students because they often had a decade of reinforced experiences that valued listening for correct answers from their teachers rather than participating in an authentic way.

To help meet these leadership challenges, CSET implemented a video study in which members of the CSET team observed videos of classroom practice, submitted by SFUSD teachers, to assess which practices related to the 5E Model were more prevalent in SFUSD classrooms and which practices were lacking or nonexistent. SFUSD and CSET then used this analysis to focus their professional learning. CSET worked alongside the SFUSD district science team to develop a professional learning arc to help teachers make sense of these shifts in teaching and learning and focus their professional learning on the areas where they needed to grow. SFUSD and CSET also intentionally set out to develop and empower science teacher leaders from across the district so they could support this major systemic change.

A Focus on Relationships to Build and Sustain Leadership

The relationship that developed between SFUSD and CSET was much more than transactional in nature. Individuals within the organizations developed authentic working relationships that built and nurtured the leadership capacity of SFUSD's team members to facilitate the changes that needed to occur. The partnership focused on leveraging the strengths and unique perspectives from both institutions to create professional learning experiences for all teachers that would have profound and meaningful impact on their classroom practices. The partnership also included CSET working alongside district and teacher leaders to develop their capacity for this new way of teaching and learning.

While CSET could supply the research and expertise for creating transformative learning experiences and building the capacity of teacher leaders, SFUSD was able to supply the on-the-ground knowledge of past professional learning, knowledge of teachers' experiences, and deep connections to the curriculum. CSET had to learn about the knowledge and past experiences of both the district science leadership team and the classroom teachers to determine how to help the team move forward with its goals. The act of co-developing the work created common purpose and understanding about science teaching and learning, and implementing the work together made that common purpose a reality. Because of the strong relationships that were built, both organizations have made efforts to explore new and interesting collaborations that support the work that began in the initial partnership.

Reflections on Leadership

As I reflect on the years I have spent focused on understanding leadership, and specifically how to help others practice leadership, there are five areas that I think are important. In this section, I describe these key areas and why they are essential.

Leadership Is a Practice

I believe leadership is a practice—something an individual does, independent of an assigned role. Leadership can be learned and must be intentionally practiced in order for a person to improve. Teacher leadership practices involve thinking through challenges, making informed and intentional decisions, and taking actions that move toward positive outcomes. Regardless of their formal leadership positions or roles, teachers can practice leadership every day in how they interact with others and contribute ideas to their professional communities. Formal leadership roles provide teachers access to resources and responsibilities that are important for distributive decision making and school functioning. However, a position does not equate to leadership. Rather, what a teacher does in a position will determine the strength of his or her leadership practice.

Leadership Is Relational

Leadership is about working alongside other people, understanding how groups of people work together, and acting in ways that help people grow. Teacher leaders need to be able to connect with others to identify what it means to influence the ways in which groups work together toward common goals. One teacher I mentored reflected on the importance of building relationships in her teacher leadership work. She said:

> I think that one of the most important parts of the work I've done this year is relationship building. I've always known that relationships are important, but I've tried to bring authenticity to all my interactions with others this year, and I hope that the time I've put in to listening to others has paid off with better relationships and a stronger network at our school site.

Another teacher leader talked about how he understood the importance of relationships only after he failed to pay attention to them as he took on a new role in leading a science department.

Practicing leadership is about developing the ability to move others by listening and really *hearing* what they need, considering different perspectives, identifying the true adaptive challenges in the group, and interacting with others in ways that honor who people are and what they bring to the table. This practice can only be done through forming authentic relationships with members of the group who are doing the work.

Teacher Leadership Must Be Grounded in Strong Instructional Practice

Teacher leaders, and specifically teacher leaders in a particular content area, need to have a clear vision of high-quality teaching and learning in their subject areas. Practicing leadership is about determining what high-quality teaching and learning could look like, considering current instructional practice, and figuring out how to move yourself and others toward that vision.

Moving toward this vision of high-quality teaching and learning also means having a clear plan about how to do so. In the multiple roles I have held focused on improving instruction, I have used different models in different settings to help achieve these goals. There is not one way to do this, and determining the plan is very much influenced by the context of the group. The common thread, however, is for teacher leaders to create a safe space for their colleagues to engage in reflective practice together and to ground that work in a common vision and plan for teaching and learning.

Leadership Must Be Nurtured

Mentoring and support are essential when it comes to developing teacher leadership. Figuring out how to practice leadership is something that happens over time and comes from trying things out and then having the opportunity to reflect on those experiences. In the beginning of this profile, I described my own leadership development and how I had mentors who worked alongside me and pushed me in ways that helped me take challenging steps. As I continue to mentor teacher leaders myself, I notice how teachers' leadership identity affects how they practice leadership. For instance, here is a comment from one teacher I mentored:

> *I am still most interested in leading as it pertains to enabling others to lead. I see myself as a teacher leader who can catalyze change but do so in a way that honors the lived experience of the adults I work with and the students I serve. I'm someone who can negotiate, mediate, and listen. I'm someone who advocates for distributed power and shared decision making and believes that communities contain the wisdom needed to solve their own problems.*

Another teacher reflected on how much she had to offer colleagues in her science department and how she feels confident enough to take up her leadership: "I now feel confident and ready to share my expertise and skills with my colleagues, thus enabling and encouraging them to reflect upon their teaching practices."

In both of these examples, the teachers talk about how they see themselves and what they can offer to others. Some teachers have been at the same school and have worked with the same administrators, departments, and teachers for years. Yet, as they practiced leadership more intentionally and developed their own sense of who they were as teacher leaders, they were able to see what they could contribute to their own learning communities. For teachers to develop their leadership identities, they need carved out time and space to reflect and talk about challenges they face with someone who will listen and push their thinking in ways that will support their leadership growth.

Leadership Is Not an Afterthought

Finally, I would like to emphasize that developing and supporting teacher leadership should not be an afterthought, but rather an important part of the design of effective professional learning organizations. In education systems, we often conflate being a "good teacher" with being a "good leader." Although being a strong classroom practitioner is certainly important, learning how to work with adults is different from working with students and requires a different kind of awareness, knowledge, and skill set. Teacher leadership needs to be intentionally developed as part of the overall investment in effective and healthy school cultures that are focused on continuous learning and growth.

RODGER'S INSIGHTS AND INTERPRETATIONS

Like many leaders, Tammy had risks and failures in her early attempts at formal leadership. The result also was typical: feelings of failure. With the support of her family, friends, and colleagues, Tammy continually sought experiences that strengthened her knowledge and skills as an educational leader. Early in her career as a middle school science teacher, she received formal recognition as a leader, which contributed to her confidence and growth.

Tammy's reflections on leadership provide several insights:

- Leadership is an active, intentional practice that extends beyond an assigned role.
- Leadership is relational, which means it requires understanding and connecting with others.
- Leadership must be grounded in strong instructional practice.
- Leadership must be nurtured.
- Leadership is not an afterthought, but instead is an important part of designing effective professional learning experiences.

13

Leading Curriculum Reform

DORA KASTEL

Dora Kastel is the director of curriculum and instruction (content area) at New Visions for Public Schools in New York, New York.

NOTE FROM RODGER

Dora Kastel is the director of curriculum instruction at New Visions for Public Schools in New York, New York. Prior to her current position, Dora was a leader of professional development programs at the American Museum of Natural History (AMNH). She began her career as a middle school science and math teacher. She earned a bachelor's degree in geology from the University of Pennsylvania and master's degrees in science and math education from Teachers College at Columbia University. She is currently working on her doctorate in science education.

I became aware of Dora's leadership while she was at AMNH, where I recognized her knowledge of science, standards, and education. More important, her leadership abilities were remarkable. She was, in my opinion, an emerging leader in science education.

By July 2019, Dora had joined New Visions and I was invited as a participant observer to a science leadership summit she was facilitating for leaders in New York schools. The context of the professional learning program was the reform of science curricula to align with the *Next Generation Science Standards* (*NGSS*) and the new science standards for New York. My aim was to observe Dora's leadership during this event.

An Early Realization of Leadership

As far back as I can remember, *leader* has been a component of my identity. My interest in leadership must have been apparent to my peers and teachers, as I have memories of being asked to lead projects and activities in Hebrew school summer camp and even when playing with friends. In elementary school, I was a table leader. In middle school, I was elected to the peer leader program. In high school, I was elected as one of four leaders on the council for our National Honor Society chapter.

"Dora is an enthusiastic student." Throughout my K–12 schooling, all of my report cards referred to my enthusiasm. I think it may have been a nice way of saying that I talked a lot; as an adult, I now recognize that so much of my learning occurred through social interaction and discourse. My friends will agree that I've always been intense and dramatic, and I'm grateful that during my adolescence I was able to channel this enthusiasm both in school and outside of school in the performing arts. In a childhood theater program, I had the opportunity to be expressive physically and verbally. We were positioned as writers, composers, choreographers, actors, and directors. Although performing arts and the nature of auditioning tend to bring out competitiveness among children, I naturally fell into a pattern supporting my peers, as I sought their support myself. The productions I worked on were ensembles in nature, and casts and crews developed strong bonds and connections. I wasn't aware that I was developing leadership skills through this participation outside of my academic endeavors, but I have found over the years that many of my skills as a science education leader can be traced back to my involvement onstage and off as a teenager.

My Experiences and Mentors

In college, rather than continue as a performer, I became a stage manager, which involved managing all aspects of huge, nearly professional-quality productions—things like coordinating actors; managing scheduling and spaces; and supporting clear communication between the director, designers, tech crew, and cast. As the stage manager, I needed to maintain strong relationships with everyone around me and ensure that all tasks were implemented successfully both leading up to and during performances. I was "in charge" in a way I had never really experienced at that scale. And nearly every aspect of this theater work prepared me to lead in the education sector later in my adult life.

One of the first things that comes to mind for me when reflecting on my professional journey, and my successes in science leadership work, is taking the advice of my mentors. I believe most of my successes can be connected to specific mentors. During high school, the mentor I identified through the science research program was Debra Curry, who worked for the New York City Department of Environmental Protection. Deb made me feel as though my high school–level contributions had true worth, and she invited me to present my research with her at the American Geophysical Union conference when I was only 17 years old. One of the many pieces of advice Deb passed along to me was to study geology in college because she felt it would be a good fit for me; even though I had a variety of other interests, I took her advice. I was surprised that I loved being a geology major, and as I became interested in geophysics, I found myself another mentor, Dr. Ed Doheny, who encouraged me to be his teaching assistant, which I ended up loving. While getting my teaching degree after college, I took a course with Dr. Michael Passow and immediately found in him another new mentor who helped instill in me the value of engaging in science learning outside the classroom. Another professor and mentor, Dr. Tara O'Neill, suggested I work at her school in East Harlem. I interviewed and accepted the job, and it was one of the best decisions I have ever made.

Once I was a teacher, Tara also suggested I participate in the Urban Advantage program at AMNH, which was an easy "yes" for me given the values instilled in me by my mentors. Working with the AMNH program, I met Dr. Jim Short. After participating in the program for a few years, I realized I wanted to lead professional learning programs. My close friend Rebecca Brian suggested that I be bold and ask Jim about this. Initially reluctant, I had enough sense to recognize that this was good advice, and I approached Jim about a job. Within two years, he offered me a full-time position at AMNH, where I began my formal work as a science education leader.

Recognize Opportunities, Build Relationships, and Learn to Facilitate

Everywhere I have worked, I have strived to take advantage of the opportunities presented to me and build relationships with the leaders of those opportunities. When I was a teacher, Tara suggested I pilot the Project-Based Inquiry Science curriculum, and I was able to meet Dr. Mary Starr, who helped me learn about the value of high-quality instructional materials. Similarly, when I became a lead teacher with the Urban Advantage program, I participated in learning from BSCS (Biological Sciences Curriculum Study) and was exposed to the skilled facilitation of Jody Bintz. Observing leaders like Jim, Mary, and Jody helped me envision the kind of leader I wanted to become. The explicit guidance from BSCS on science leadership was helpful for me in developing skills to work with adult learners—teachers within my school and those who attended workshops at AMNH. As I was transitioning from classroom teacher to full-time professional development provider, I attended a teacher workshop given by Dr. David Randle and Dr. Cristina Trowbridge on how to use data to teach about climate change. Even though I was a participant, I decided to pay attention to all of the facilitation moves they were making and took copious notes. I asked to meet with Dave after the workshop and asked him questions about the rationale behind his strategies.

While working with Jim at AMNH, and with his encouragement, I worked on numerous projects—starting with small ones on a local scale and working my way up to national-scale projects that allowed me to enact my visions as a leader. My confidence in presenting work at conferences can certainly be traced back to my high school experience with Deb, as well as with Jay Holmes, a colleague at AMNH who was another strong role model in terms of learning how to support the growth of teacher leaders.

Leadership at New Visions for Public Schools

As an associate director at New Visions, my work involves creating a coherent system of support to ensure teachers are able to use our materials effectively. For this work to be successful, we have to think about leadership at many different levels. One key component of our work has been supporting leaders who are in positions to help the teachers across our network. Assistant principals who supervise science teachers, science department chairs, and district instructional leaders are all positioned to positively influence the successful implementation of our work across schools. Our professional learning includes a strand of curriculum-based professional learning for teachers and a strand of professional learning for leaders supporting teachers using our curriculum.

Here are several actions that have been essential in planning professional learning:

- Before anything, set goals.

- Recruit leaders at different levels.

- Provide multiple opportunities for support and relationship building.

- Anchor the professional learning experience in instructional materials.

- Build facilitators' capacity to adapt and lead other adult learners.

- Articulate a common vision and share concrete action plans.

And here are a few things I have learned about leadership along the way:

- See yourself as a leader. Begin leading when you have an opportunity.

- Stay close to the students, teachers, and schools your leadership affects.

- Keep learning, especially from your mentors. Build relationships and a network of support.

- Create the change you want to see.

- Facilitate the development of new leaders.

RODGER'S INSIGHTS AND INTERPRETATIONS

I gained insight into Dora's leadership when I observed her at the New Visions Science Leadership Summit in July 2019. My recognition of her leadership was greatly augmented by observing her in a context of providing professional learning for science teachers, which I describe below.

New Visions is in a multiyear project to update current curricula to align with the New York state science standards. Dora and her team worked alongside teachers, school leaders, and district leaders to develop a vision for a curriculum aligned with these standards.

Equity was an important goal, as New Visions believes that all students deserve to engage in authentic science experiences, and the changes implied by the new science standards can be achieved through the leadership of science teachers.

As I participated in this workshop, what did I observe of Dora's leadership? Dora was clear about the purpose and plans as ways the program would support the participants' understanding of the implied changes of the science standards. Every effort was made to establish community among participants as Dora answered questions large (e.g., What are the national standards?) and concrete (e.g., What do the standards mean for my grade or course?). As participants were immersed in an activity of curriculum reform, Dora circulated among participants, listened to concerns, responded to questions, and, most important, treated all individuals with dignity and integrity. These are all important skills and qualities for a strong leader, and Dora exemplifies them well.

14

Six Stories From a Leadership Journey

JOHN P. SPIEGEL

John P. Spiegel is the director of curriculum and instruction at the San Diego County Office of Education in San Diego, California.

NOTE FROM RODGER

Beginning in 2003, I observed John Spiegel's development as a leader as he transitioned from his role as a classroom teacher to principal and eventually the director of curriculum and instruction at the San Diego County Office of Education. He has continually demonstrated leadership in science education.

John has a bachelor's degree in physics from San Diego State University and a master's degree in educational leadership from the University of San Diego.

Introduction

I didn't consider myself a leader when I began my career in education. I just wanted to teach science. After getting my degree in physics and completing my credential program, I set my sights on becoming a high school physics teacher. When I was hired by a school district, however, the district offered me a position teaching science in a middle school. I quickly learned that thriving in that environment required the ability to love your students more than the content.

While I was settling into the groove of teaching, I was fortunate to be mentored by others who saw potential in me. They invited me to serve on committees and take on additional responsibilities that helped me grow my confidence and expand my leadership skills. Over time, I transitioned from being a teacher in a classroom to an administrator at a school. I began to see opportunities to make larger contributions to the education system beyond the direct influence I sought to have on my students. Since then, I have continually tried to develop and refine a style of leadership that influences, supports, and makes a difference to others, regardless of the position or title I might have.

I will share some personal stories from different moments in my career that define me as a person and a leader. I tell each story as if it were happening in the present moment. I will then offer reflections that connect the stories.

My First Day of Teaching

As I sit at my desk in the late afternoon on an early September day, I am physically exhausted, yet my mind is still racing as I think back on my first day of teaching. A couple hours ago, I was standing in front of students leading a demonstration while they recorded predictions and observations. The demonstration involved a two-liter bottle filled with water. On the side of the bottle were three taped-up holes that prevented water from leaking. I had asked students to predict what they thought would happen when the tape was removed from the top hole. When they finished their predictions, I removed the tape and, to the surprise of many students, no water came out. I then asked them to predict what they thought would happen when the tape was removed from the middle hole. One student shouted that he didn't think water would come out again. I invited the student to the front of the classroom and asked if he was confident enough in his prediction to sit in a chair while I held the bottle over his head and removed the tape. He excitedly sat down, and all of the students were on the edge of their seats waiting to see what would happen. I stood up on another chair and held the bottle high over his head.

At that exact moment, the vice principal walked into the classroom to observe. My heart started racing, and I suddenly felt nervous. My mind raced back through all of my preservice coursework as I tried to figure out what to do when an administrator comes into the classroom on the very first day of your teaching career. I felt all my confidence slip away for a few seconds until I thought about how much time in my science methods course was focused on helping me figure out what kind of a teacher I wanted to be.

I looked out at the students who were eagerly anticipating what would happen when the tape was removed. I paused a little longer and quickly pulled off the tape. Water streamed from the hole in the bottle and spilled out on the head of the student below. He jumped up and students began laughing. The vice principal smiled. I felt exhilarated, and my passion for science education was born.

Physics First

I am looking across a room filled with district and school leaders from around the country at the beginning of the BSCS (Biological Sciences Curriculum Study) National Academy for Curriculum Leadership. As I listen to people introduce themselves and share their experiences in education, I feel out of place. I am still in my first few years of teaching and continually trying to figure things out in my classroom. Because I recently started teaching a freshman physics course at my school, the central office science director thought I might be a good addition to our district team.

The team assembled from my district consists of teachers, site administrators, and district leaders. It is a powerhouse group of diverse perspectives, backgrounds, and personalities. We have come together to establish a vision and implementation plan for high school science reform in our district, where the graduation requirement in science will be increased from two to three years and all non–college prepa-ratory science courses will be eliminated. The traditional science sequence will also be inverted, as all students will take physics in 9th grade, chemistry in 10th grade, and biology in 11th grade.

I am assigned to a small team tasked with designing and leading professional development for teachers across the district. Because physics has been moved to 9th grade, the number of credentialed physics teachers needed increases significantly, requiring many life science teachers to secure a secondary authorization on their credential. The professional development our team provides focuses on building physics content understanding as well as inquiry pedagogy to support the learning of all students. Because of this effort, I begin to split my time between teaching in the classroom and working as a resource teacher at the district level.

Over the next several years, I learn the value of vision as our team continues to support physics teachers. At the core of this vision is a belief that all students should have access to physics and all students can be successful if they are provided sufficient support. These beliefs are regularly challenged in professional development sessions. I am having my first experiences with dealing with resistance to change within a system.

Small-School Reform

I am a principal standing on the stage at the commencement ceremony. The pride I feel shows across my face as I look at each of the graduating high school seniors, hand them their diplomas, congratulate them, and shake their hands. These are the first graduates of the Invention and Design Educational Academy, one of four autonomous small schools on the campus of a former comprehensive high school.

The journey to this moment has taken three years and began while I was a new administrator at the former high school. Despite the incredible efforts of many dedicated staff, the organizational structures and curriculum of the school were failing to meet the needs of the diverse student body. With an infusion of financial support, a team of teachers, administrators, and parents began a process of reimagining the high school experience so it would have a greater focus on building stronger relationships with students, providing curriculum more relevant to learners, and raising expectations for students.

After a year of initial study, four design teams were created to develop new autonomous schools that would share the same campus. I was tasked with co-leading one of the small-school design teams. Student surveys and focus groups were used to shape the emerging concept of a school with an engineering theme. We developed an advisory program and created a yearly exhibition for students to highlight their accomplishments and successes. Halfway through the planning year, I was selected as the principal of the new school. The to-do list was endless, and there were not enough hours in the day to get everything done. Facilities needed to be sorted out, staff needed to be hired, the master schedule needed to be built, and recruitment and marketing needed to happen. In addition, given that my school was one of four on the same campus, shared-space agreements needed to be written, supervision schedules needed to be created, sports and common clubs needed attention, and on and on.

After the school opened, new challenges and problems arose. Student attendance was still low and new procedures had to be figured out and student supports implemented. Adults on the campus did not have consistent expectations for all students, which created tensions among staff and students. We felt increased pressure from the district to show immediate results. The hours were endless but also incredibly satisfying.

Through all of these challenges, my personal and professional growth continued to evolve as I began to learn how to delegate, problem solve, help groups reach consensus, and adapt to a constantly changing environment. As a leader, I began to understand that structural change in a school is easier than cultural change.

District Reorganization

I am now the district curriculum leader in science, walking back to my office after a meeting with the deputy superintendent. My mind is reflecting on the conversation that has just occurred. A couple of weeks ago, I was tasked with developing a proposal to cut costs for the district's Science Refurbishment Center. In the meeting that just concluded, I presented several models to save money by reorganizing the support and services the center could provide to schools. The reaction of the deputy superintendent was not at all what I expected. After listening to my presentation, he explained that he didn't want me to find a way to save money, but rather he wanted me to devise a plan to close the center entirely. I was, quite frankly, shocked and told him I couldn't do such a thing.

A little background might be helpful here. The opening of the Science Refurbishment Center in the district was a key strategy in science reform and support for elementary schools. The adopted elementary curriculum included kit-based science consumable materials that needed to be replenished each year. Prior to the Refurbishment Center's existence, access to such science materials was inconsistent across schools and created inequities in the quality of science learning experiences for students. In some schools, kits had been dismantled entirely or were missing materials, making it impossible to teach the science curriculum. The Science Refurbishment Center—along with ongoing, intensive professional learning for teachers—contributed to the steady growth in student achievement in science in the district.

When I was hired as the district curriculum leader in science, the Science Refurbishment Center was already in place. One of the first responsibilities I had was overseeing the process for selecting new instructional materials for elementary schools. When an updated version of the same kit-based curriculum was selected, the district science department leveraged the refurbishment system in place to ensure teachers and schools would continue to receive support throughout the life of the curriculum.

In the two years I had been at the central office, the district underwent several changes in leadership, including two superintendents. When asked by the new deputy superintendent to come up with a plan to close the Science Refurbishment Center, I saw it as the undoing of many years of science reform efforts and the beginning of a de-emphasis on science by the district. I told the deputy superintendent in no uncertain terms that I would not design a plan I could not support and that I believed the steady growth in student achievement in science would cease without the support of the Refurbishment Center.

As I walked back to my office after the meeting, I found myself questioning what I said, how I said it, what would happen next, whether I was right or wrong, and even if I would still have a job. At the same time, I felt a sense of resolve in myself that whatever the outcome, I had acted in what I believed was right and in the best interest for science learning in elementary schools.

Returning to Science

I am a principal again, catching up on e-mail in my office when I read something I think may change my career path yet again. I have received an e-mail announcement for a science coordinator position at the San Diego County Office of Education. Among the listed duties for this position is the responsibility to help lead the implementation of the newly released *Next Generation Science Standards*

(*NGSS*) across the 43 districts in the county. I close my laptop and take a walk around the school to think.

I have been at my school for almost four years. The school serves kindergarten through eighth grade. I have enjoyed seeing many of the students grow up. When parents ask me if I have children, I reply, "Yes, I have 575 children, and I love them all." The days are long and the calendar is always active. I arrive at school by 6 a.m. most days, create my copious to-do list for the day, and almost never get to many of the tasks, as the moment-to-moment daily events always prove to be more urgent. I often find myself struggling with trying to find a balance between leading the instructional program of the school and managing the overall operations.

As I walk the hallways on this particular day, I occasionally stop and look into a few elementary classrooms. I observe students interacting both with each other and with their teachers. At one point, a student approaches me and we have a short conversation. We bump fists before she rushes back to her class. As I near the middle school area, the passing period begins. I watch students move from one classroom to another.

I am pondering the possible new career opportunity as I walk back to my office. I feel a heaviness settle over me as I realize I have a choice to make between two things I love. Leading a school is as challenging as it is rewarding. However, the e-mail I read earlier surfaced something in me that I have been feeling for some time but was afraid to fully admit. I chose this career because of my love of science education. My leadership journey has led me down various paths that helped me grow and gain experience across the education system. Now I am recognizing that I have been truly missing the science aspect of it all.

I go home that night and excitedly begin filling out the application.

Next Generation Science Standards

Today is my eighth year as coordinator and director at the San Diego County Office of Education (SDCOE). I am thinking back on the many experiences I have created and been part of as I have helped districts and educators understand and implement the *NGSS*. One moment in particular stands out as I reflect on the past several years and helps me imagine what the next decade might hold for me.

A few weeks after I was hired, a local news station contacted SDCOE to do a television and radio interview with someone about STEM education and the new science standards California was in the process of adopting. The communications department at SDCOE reached out to me and asked if I would participate in the interview. I felt terrified but reluctantly agreed.

I spent the next several days trying to figure out answers to possible questions that might be asked: What are the new standards? How are they different and improved from the current standards? How will science look and feel different for students and teachers? My mind raced as I imagined being asked questions that I didn't know how to answer. I felt somewhat afraid and wanted to do my best during the interview. The day of the interview came and went in a blur. I don't remember what was asked or my responses. I do remember the personal and professional commitment I made to myself afterward to learn as much as I could about how to best implement the *NGSS* across the county and state. I became determined to get involved in leadership committees, collaborate with as many people as possible, and become an asset for educators to help improve science education for all students.

Today, similar questions and new ones often run through my mind. Figuring out the answers keeps me excited about coming to work each day. I have a much better understanding of my role as a leader now as well as how I want to continue to lead. Eight years ago, I was afraid of questions I could not answer. Now I understand that leadership is the willingness to ask such questions, then try to find the answers.

Some Personal Reflections on My Leadership Journey

The challenge of trying to describe my leadership journey is that my perspectives on it change from time to time. The stories I shared here are defining moments for me. Next, I offer some thoughts on leadership that come from the stories I have shared and others like them.

Across my career, I have learned to recognize the value of my relationships with mentors who inspired me, helped me set and achieve goals for my personal growth, and encouraged me along the way. One mentor saw my potential and opened a door for me to learn that helping adults learn and grow is as satisfying as teaching children. Another supported me as a new administrator by helping me think through situations and problems. A third helped me create realistic action plans for focused goals.

I have always felt a tension between my confidence to lead and my doubts about my ability to do so. Early in my career, I often saw this tension as a weakness, but I now realize how much of a strength it is. My confidence allows me to take risks and step into new opportunities with excitement. My doubts, or moments when I question my actions or beliefs, keep me grounded and eager to reach out to others for support. They also provide me chances to step back, re-evaluate my actions, and change course.

I have learned as a leader how to step outside the box of the way things currently are and ask these questions: What if we do this? Why don't we try this? Many of the initiatives I shared in my stories—such as the physics-first movement, the small-school reform, and the Science Refurbishment Center—are no longer in place; some might say they were failed efforts. Each of them, however, involved people rethinking what is possible to better the education system. These efforts shifted belief systems about how to improve learning opportunities for students. For that reason, and for how they touched students' lives, I think these efforts achieved success.

Leadership involves the willingness to step in, make a contribution to a cause, and serve others. It is about not being afraid to work hard. A leader needs to be able to adapt and respond to change. Leaders build doors for others to walk through to create something better. Sometimes others built doors for me, and I hope I am continually building doors for others as well.

Finally, leadership is about finding a passion and being inspired by it. My passion in education has always been science. I love learning about science, teaching science, seeing students be curious about the science world around them, inspiring and helping science teachers, and advocating for quality science experiences for every student in every classroom in every school. This passion is where my leadership story begins and is, fortunately, how it continues today. In between the beginning and now were many opportunities that took me away from that passion but served me well by providing needed growth and experience.

RODGER'S INSIGHTS AND INTERPRETATIONS

In June 2019, I attended a workshop led by John and his team, including Chelsea Cochrane, Chrystal Howe, and Kathy Bowman. The workshop was for teams of two or three from nine school districts in San Diego County. The program addressed evaluation of instructional materials, and my purpose was to observe John and his colleagues as they developed leadership knowledge and abilities for the participating teams.

Activities during the two days included analyzing instructional materials, discussing the design of units, addressing the crucial features of *NGSS*, and aligning lessons with *NGSS*.

John and his team demonstrated leadership abilities such as listening for and responding to the concerns expressed by individuals and teams, modeling the knowledge required to conduct a workshop, engaging teams in the actual activities they would conduct, clarifying the vision underlying the evaluation, and answering questions about implementing the process for evaluating instructional materials aligned with *NGSS*.

John and his team demonstrated many of the personal or person-centered aspects of leadership as each team member interacted with participants and made presentations. Specifically related to John, it was clear his team had internalized several of the key themes of leadership as he has described them in his profile.

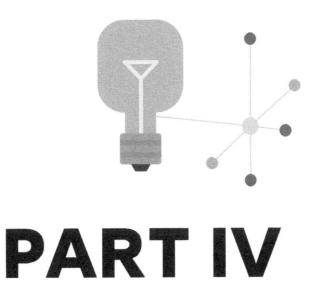

PART IV

Some Leaders Support Science Teachers by Guiding Professional Organizations

15.

Beyond the Classroom

Serving All Students

STEPHEN L. PRUITT

Stephen L. Pruitt is the president of the Southern Regional Education Board in Atlanta, Georgia.

NOTE FROM RODGER

Although I had met Stephen Pruitt years ago, most likely at an NSTA conference, I came to realize his leadership during our work on *A Framework for K–12 Science Education* (NRC 2012). Both Stephen and I contributed to that NRC report, which was the foundation for the *Next Generation Science Standards* (*NGSS*; NGSS Lead States 2013). Subsequently, Stephen went on to direct the development of the *NGSS*. He eventually joined me in writing *Perspectives on Science Education: A Leadership Seminar* (Bybee and Pruitt 2017).

Stephen began his career as a high school chemistry teacher in Fayette County, Georgia. In July 2018, Stephen became the sixth president of the Southern Regional Education Board (SREB) in Atlanta, Georgia. Before going to SREB, Stephen was Kentucky's state commissioner of education, and he has assumed several leadership roles for the Georgia Department of Education.

Stephen has also served as a board member for the Council of Chief State School Officers, president of the Council of State Science Supervisors, and a member of the writing team for the Science College Board Standards for College Success.

A native of Georgia, Stephen holds a bachelor's degree in chemistry from North Georgia College, a master's degree in science education from the University of West Georgia, and a doctorate in chemistry education from Auburn University.

Moving Beyond the Classroom

First and foremost, I spent 12 years as a high school science teacher. I am very proud of my time in the classroom. I left the classroom not for money or titles, but rather because I felt I could support more students by working with teachers.

In my previous roles in state government and regional and national organizations, I had opportunities to manage educational change and lead people. When I began work at the Georgia Department of Education (GaDOE), science education in the state was not serving all students. In 2003, only 36% of African American students passed the Georgia High School Graduation Test in science; those students who did not pass were unable to receive a high school diploma. We instituted programs meant to increase the success of students in science, including by implementing the largest change to the state standards in Georgia's history and a program to support science teachers in struggling schools. As a result, the state passage rate for all students went from 68% to 90% during my tenure at the agency, and passage rates for African American students went up to 84% (Georgia Department of Education 2002, 2010). In addition, during my time as curriculum director at GaDOE, I oversaw the implementation of some of the largest budget cuts to education in Georgia's history. These changes required a solid knowledge of state and federal regulations and the creativity to help districts and the agency literally do more with less funding. I was also responsible for moving the strategic plan of the agency forward and focused on ensuring that all efforts supported the organization's goal of leading the nation in improving student achievement.

As chief of staff, I coordinated and worked with our state board of education, including answering board members' questions and preparing them for meetings. I always felt that some of the most important tasks involved working with the board. The more I worked with them, the more smoothly our board meetings proceeded. I was often able to deal with issues long before they became public or needed to be addressed in the actual board meetings. They trusted me, and this trust created a great working relationship. These experiences shaped my understanding of the need to work with diverse groups to develop a common vision and goal for the whole enterprise of education.

It is my fervent belief that people within an organization can contribute if they are put in the correct positions and given appropriate direction and support. In my role at Achieve, I led one of the biggest reform efforts in science education in the early 21st century. The development of the *NGSS* required me to understand not only the science and science education research needed for new standards but also the politics around implementing the standards. I worked with many groups and individuals to build the best set of standards possible. The ability to operate in a political environment and rally support from associated organizations and partners is critical in an era of standards-based reform. In the development of the *NGSS*, I had the honor of working with the best and brightest in science, science education, and policy from around the country. How did this shape my perspective? Perhaps the most significant insight I gained was the irrefutable reality that leadership matters. The leadership provided by the teachers, administrators, science researchers, policy makers, and engineers on this project was an incredible example of what can happen when there is a common goal—in this case, to increase access to a quality science education that serves all students.

In October 2015, I became the commissioner of education for the Commonwealth of Kentucky. While leaving full-time science work was tough, the opportunity to work in the state of Kentucky was too great to pass up. My role allowed me to work with all aspects of education, and my experiences with the *NGSS* and Georgia's education system have given me a unique perspective. Joining the Southern Regional Education Board has enabled me to expand the geographic reach of my goal of serving all students.

Before moving on from my professional experiences with leadership, I will make a point about the critical leadership of science teachers. I am often dismayed, and even angered, when I hear one of our wonderful science teachers say, "I am *just* a teacher." We do our colleagues, our students, and ourselves a tremendous disservice when we belittle ourselves this way. I think this has become a societal label about us, and if it is ever to stop, we educators must stop it. Even if the word *just* is what the world wants to use to label us, we perpetuate that perspective by using this word ourselves. Educators are the most passionate, compassionate, caring, and intelligent people I know. What is missing, unfortunately, is leadership outside our classrooms or buildings.

Reflections on Leadership

How we lead is up to each of us individually and must fit our personalities. I would also say, however, that we may not know our style if we do not spend some time in the study of other leaders and how they are looked at by those they lead. I have observed many leaders, studied them, and tried to emulate the qualities that fit my personality. In this profile, I will discuss the leaders who have affected me, what I learned from them, and how their leadership affects my daily decisions and actions. Some of these leaders are people I have had the opportunity to know personally and some are historical figures, while others are, oddly enough, fictitious, though I still learned from them.

In *Harry Potter and the Deathly Hallows* (Rowling 2007), Professor Dumbledore says, "It is a curious thing, Harry, but perhaps those who are best suited to power are those who have never sought it. Those who, like you, have leadership thrust upon them, and take up the mantle because they must, and find to their own surprise that they wear it well." Leadership is not something science teachers look for, but we are in a time when educators are having leadership thrust upon them. We wear it well, but we must still choose to put it on. The traits we apply in our classrooms every day are the same ones that make us good leaders.

Presence and Respect

The first trait of a good leader is presence. You might say, "Duh." When I say *presence*, I mean that when a leader is in the room, people know it and respect the leader. This does not mean they blindly follow him or her, but they recognize there is someone who may have achieved much, yet is eager to work with others so they achieve more together. I have seen this trait in some of the greatest leaders I have known or read about. To be clear, this is not the person who says, "Here I am, look at me!" Rather, it is the person who need not say a word, yet people know he or she is there. I am a fan of George Washington. I love reading historical novels, not just because I am a history buff, but because of what I can learn about leadership. It is clear why Washington holds the station he does in our country. In many historical novels and accounts, you will find that Washington did not give fiery speeches, nor was he always addressing the crowd. More often than not, he simply had a presence in the room or talked to small groups. Everyone knew who he was and what he was about. He was this way his whole life. He knew the value of good relationships and how to hold people accountable. I wanted to build my knowledge and my abilities as a leader and leverage my relationship skills to build presence. I do not have a need to be in front all the time, although I am comfortable there. But I want to lead by example and offer my thoughts on tough problems because I believe I have something to offer.

Another great example of presence is my other favorite president, Abraham Lincoln. We know his speeches and achievements. What is often left out of his story is that he was a fairly soft-spoken man with a high voice. He, like Washington, did not feel the need to "control" a room. He did anyway, but in a different manner. He would lead through his stories, for instance. If a discussion was off track or moving in a direction he did not like, rather than always redirecting it with a mandate, he did so by sharing one of his many stories. Lincoln was also a master at positioning himself in a room full of people. His presence was felt, but he often decided where to stand or how he could empower or comfort someone based on his place in the room. Like a good classroom teacher, he understood that where you stand and how you move around the room make a difference.

If educators build presence, they will no longer say, "I am *just* a science teacher"; they would simply say, "I am a science teacher," and that will be enough to drive the conversation. During the development of the *NGSS*, I knew I needed to play a prominent role, but I also needed to not be seen as the leader. There were many smarter and more savvy people I had the chance to work with on this team, but I like to think there was some comfort in knowing I was there. Influencing others with your presence is a goal I will continue to pursue. A leader is not the loudest or even the smartest person but the person whose leadership can influence others to be better than they thought they could be.

Courage and Patience

A second major trait for a good leader is courage, complemented with patience. This trait might seem obvious, but I suggest it is not. When we were developing the *NGSS*, I often said that we must have the courage to be patient. In education, we are all about doing things quickly, and doing so often fails us. After all, no great education initiative ever failed in the vision phase; failure happens in implementation. It is here we need courage. It is courageous to be out in front of the development of an idea and have the patience to plan, envision, build relationships, execute a difficult aspect of a vision, and think outside the box on a tough issue. The *NGSS* are very different than any state standards we have ever had, and it has taken courageous teachers to implement them in classrooms when they realize the type of curriculum and instruction required are very different than what they have done in the past. We have also needed to have courage and patience so as not to rush into testing or accountability with the *NGSS*. Assessment of these standards is different than for previous standards because the aim was to push change in our science classrooms. But rather than rushing, most have been patient, despite criticism from those who do not understand the need to get it right. So, I would say courage is actually doing what you know to be right, even when others around you are telling you to go in a different direction. This leads me to my next point about courage: the need to do what is right, not what is easy.

An unfortunate part of our profession is that it is common to feel cut off from the outside world while we are in our classrooms working with students. It is easy to become cynical about change. A courageous educational leader realizes we can always improve. Achieving educational excellence is not the destination; it is the journey. Educational excellence today looks very different than it did 20 years ago and different from what it will be 20 years from now. I am inspired by the classroom teachers courageous enough to take risks and do what they know they must to improve their instruction for the sake of their students. They do this knowing they are learning themselves and that any change or transition is not easy. As Professor Dumbledore also said, "Dark times lie ahead of us, and there will be a time when we must choose between what is easy and what is right" (Rowling 2000). For some, new standards meant dark times. I am happy to say that I believe for most of us, though, they represented new hope. I have seen some incredibly courageous teachers who have said they simply won't stand for dismissive views. They push forward, serve on state and local committees to get the implementation right, and stand as models for the rest of us who struggle each day. They did not take the easy way; they went the right way.

Perseverance and Relationships

Washington and Lincoln certainly went the right way, and that way was never easy. What they showed us was that perseverance and relationships matter. Washington showed perseverance in his military battles and his early struggles with trying to move up the ranks. He built relationships that eventually led to him owning his beloved Mount Vernon. He had presence because he had perseverance and relationships.

Lincoln certainly had the same characteristics. The 13th Amendment, which was incredibly controversial, passed in part due to his perseverance and personal relationships. He showed people he cared in small ways, such as placing a hand on a shoulder when his men were panicking or taking the time to sit and talk about an issue with people regardless of their societal level—these were ways he built relationships. He lost some relationships with the passage of the 13th Amendment, but he also knew passing it was the right thing to do. It may appear odd to place these two traits in the same section, but they are connected. A person can persevere through adversity, but an idea can only persevere if many support it. This support is built through relationships.

The development of the *NGSS* required perseverance and relationships. I was able to build many during that time, and it's possible I hurt some relationships as well. We needed to persevere even when other standards were under attack. I felt the future education of our country was at risk. The relationships between key partners and educators were important to the standards themselves persevering, so it was essential that leaders from all fields came together to work on and advocate for the standards.

Vision and Context

As an education leader, one must realize the importance of having a vision. Without vision, there is nowhere to go. I have had opportunities to work with some leaders who were great because they had vision. They were not only out-of-the-box thinkers—I am not sure they even knew where the box was. I think of Wayne Robinson, the principal of the first school where I taught and one of the greatest education leaders I have known. He had all the traits I have mentioned and more. But what I remember most about him is his vision. He was committed to making our school the best in the county and probably the state. He did so with a passion and presence that still inspire me today. Many times people will talk about great visionaries as being just that and not concerning themselves with the details. Wayne taught me to "sweat the small stuff" because that is what will undermine your work. He kept his eye on the big vision while making sure nothing derailed it. At the end of the day, his vision for *why* he wanted the best school in the state was clear—our students.

Vision without context is a dream quickly forgotten; vision with context is a dream come true. As education leaders, we should always be an advocate for education. However, as soon as that advocacy becomes about our content more than our students, or about preserving a way of teaching instead of changing the lives of students, we have lost.

Equity and Expectations

I had no plans to leave the classroom. I loved, and still love, everything to do with being a classroom teacher. At the core, while I loved science, my belief that I could make students' lives better drew me to a broader career in education. As my perspective grew beyond the classroom, I saw how my expectations did not match those of others. At that point, the district I taught in still had levels for our high school classes. I began to notice I had different expectations in my classroom depending on the "level" I was teaching. I thought I was making decisions for the benefit of my students, but I realized that my students would do what I expected them to do. I also noticed that some things were more important than others, and, oddly enough, they were concepts from the contemporary standards.

When I moved to the GaDOE as the science supervisor, the inequity of our science expectations became more real than ever before. As I did my own research into the reason for this inequity, I found that students who did not perform well on the exam also had not been exposed to a rigorous set of science coursework. This finding was my first experience with the phenomena that I have come to call the "less is non phenomena," which means that we in the education community have the expectation that students who do not show success in science and mathematics early and often will somehow get better at these subjects by doing less work in them. Students who do not perform well in science and math take fewer courses in the subjects, have lower expectations placed on them, and continue to perform at lower levels than students who have higher expectations placed on them. The only thing these students get more of is remediation, so we give students who have come to dislike the subject more "drill and kill" to prepare for an exam, and lower expectations equal little or no actual improvement in performance. As a result, students who are not given the opportunity to perform at high levels lack the knowledge and abilities for success in the workforce after they graduate—if they graduate! This is, from my perspective, a grand challenge.

A Few Final Thoughts

I would like to reiterate that the critical issue is equity. From the beginning of the development of *A Framework for K–12 Science Education*, the issue of scientific literacy was always viewed through the lens of equity. It is important in the 21st century for all students to be prepared for the life they choose. As we developed the *NGSS*, it was not an accident that our focus was *all standards, all students*.

The standards have the ability to influence the equilibrium that exists in contemporary education. As with all systems at equilibrium, education will only change with stress. The *NGSS* were meant to be that stress. The *NGSS* should affect other components of the education system, such as education policy, instruction, and assessment. If this reform is to come to fruition, all of these areas must be viewed through the lens of the *NGSS*. All major components of the education system will need to change to ensure that these standards work for all students. Only if this change occurs will all students have the same opportunities for success beyond school.

I will end with a short commentary on leadership by science teachers. I mentioned earlier that we cannot continue to say, "I am *just* a teacher." We need to use our teacher voice for something more than getting students to be quiet. We need to proudly and boldly proclaim, "I *am* a teacher!" As education continues to become more political, our profession must stand ready to be a leader in our country. Through strength, presence, perseverance, relationships, vision, context, humbleness, and honor, we can be a voice for students. In fact, we know from research that important voices in our communities come from teachers. My hope is that someday leadership for our educators will be the norm in professional programs and that our teachers will stand with one voice to say we are the experts in education. We *are* teachers.

RODGER'S INSIGHTS AND INTERPRETATIONS

Stephen clarifies the theme "leadership matters" in his profile and focuses on the need for presence, respect, courage, patience, perseverance, relationships, vision, context, equity, and expectations. Stephen's profile weaves insights and understandings that should help science teachers influence others to be better than they thought they could be.

Stephen's examples show the balance of having clear goals and the means of achieving those goals. Let me conclude with observations of times when Stephen led by example.

For more than a decade, I have had numerous opportunities to observe Stephen's leadership. In many cases, I was "in the room," while in other situations I witnessed his leadership from a distance. I saw in his leadership a broad and deep understanding of issues, as expressed in his answers to questions. He showed clear perceptions of the reality of situations related to national or state policies, issues related to curriculum materials and assessment, and problems around reforming classroom practices. He modeled traits and themes he discussed in his profile, such as presence, courage, perseverance, and equity. He shared his recommendations in everyday, understandable language, including his preference that teachers not say, "I am *just* a teacher." Finally, he has demonstrated a calm and steady demeanor in difficult situations.

Stephen has also experienced the unfortunate side of leadership. As you achieve your goals as a leader, you likely will be subject to criticism, some of which will be unfair. I know of two instances where he experienced this aspect of leadership, and in both cases he demonstrated self-restraint and self-confidence. I will use Lincoln's example of how to deal with unjust situations, which can be summed up as needing to have courage and maintain a sense of humor (Phillips 1992).

16

Led to Lead

PETER J. MCLAREN

Peter J. McLaren is the executive director of Next Gen Education, LLC, in North Kingston, Rhode Island.

NOTE FROM RODGER

My first meeting with Peter McLaren was in September 2010, when he was the science and technology specialist at the Rhode Island Department of Education and president of the Council of State Science Supervisors. He now works as a consultant for states, districts, and organizations to improve school science programs.

Peter served on the national writing committee for the *Next Generation Science Standards* (*NGSS*), the National Academy of Engineering's Guiding Implementation of K–12 Engineering Committee, and the National Academy of Science's committee for developing assessments for the *NGSS*.

Peter graduated from the University of Rhode Island with bachelor's and master's degrees. He was a teacher of science for 13 years. He was recognized with the Milken Family Foundation National Educator Award in 2001 and as the Rhode Island Science Teacher of the Year in 1995.

What I Learned From My Father

When I turned 16, I experienced a rite of passage: getting my first job. After school and on Saturdays, I was a part-time clerk in a supermarket, where I worked in the produce department, loaded milk and dairy products, and operated a cash register. What made my first job truly unique, however, was that my father, Charlie McLaren, was the meat department manager at the same supermarket. Listening to and learning from my father were the first experiences that led me to lead in the future.

Working in the same building as my father was an eye-opening experience. Though I could not work directly with him (you had to be 18 to work in the meat-cutting environment), I became aware of a whole different side of my father. For the first time, I was truly able to see what he did every day when he left for work. He worked incredibly hard, and I later associated this with the idea of having a work ethic. His job was incredibly physical, involving breaking down 500 lb. sides or 200 lb. hindquarters of beef into the products for the market's customers, all while working in a refrigerated environment. Despite his challenging job, my father maintained a wonderful sense of humor, whereas others might go about this type of work grousing about the conditions. His work ethic and humor positively influenced the people who worked directly with him in the meat department, as the other cutters and counter clerks seemed to enjoy their work. The longer I watched my dad, the more I was able to see that the positive work environment in his department emanated from him. He set the tone and made sure all of the aspects of the day-to-day operation of the department were taken care of with little stress.

NATIONAL SCIENCE TEACHING ASSOCIATION

You might be asking yourself what my recounting of working with my dad as a 16-year-old has to do with my leadership. When I was asked to share my experiences with and perceptions of leadership, I realized that Charlie McLaren was really my first example of a good leader. Leadership, in the broadest sense, is the ability to influence others in order to accomplish a set of goals, and Charlie certainly fit that definition. He was always ready to share a joke with the people in his department to loosen them up, yet he led by making sure the customers and production were always the priority. Playing off of his demeanor and work ethic, the personnel in his department followed suit, and many customers came into our store simply to buy meat. His department was that good.

I have to admit I was jealous of the people who worked in my dad's department. My supervisor was all business and always checking up on me and critiquing my work. I suppose that because I was a young lad, new to the world of work, he had to keep an eye on me. When I complained privately to my dad about my boss's management style, he said, "You can learn as much from a bad manager as you can from a good manager." I still use that quote to this day.

One night, while having dinner, I asked my dad why everyone seemed to be enjoying themselves in his department. He said, "No one ever said that you have to be miserable when you work hard. People work better when they are happy, and I try to make them smile when I can." He knew that humor relaxes people and relieves stress.

Charlie had an uncanny knack for making people understand that no matter what they do, they are an important part of the business. He once told me, when I complained I had to go out and collect shopping carts in the parking lot, "If the least significant task is not accomplished, then every other aspect of the business is affected. If a customer cannot find a parking space because there are carts all over the parking lot, then they will probably go to another market." That really hit home for me. I perceived that what I was doing was a menial task, but my dad gave me the perspective that what I was doing was an important aspect of our business.

Applying Lessons Learned to My Science Classroom

As I progressed throughout the various jobs I have enjoyed in my career, the lessons I learned from my dad influenced me greatly: Work hard but laugh harder. Lead by example. Make sure people understand that they matter, no matter their role.

I eventually left the supermarket to become an eighth-grade science teacher. I found that I used many of my dad's lessons in my classroom. I became known as the "mad professor" in my school, using humor to get kids laughing, reduce their stress levels, and, consequently, prepare them to learn.

During the initial years of my teaching career, I began hearing the term *teacher leader*. I had a hard time understanding exactly what the term meant. After all, we teach students, but they are not subordinates or employees in the context in which I held leadership. I associated leadership with business or the military, where there was a hierarchy. As a new teacher, I was always trying to learn new and better ways to reach my students, but I found also myself working in a closed-off, autonomous environment, so I was not able to enjoy examples of best practices in action. I asked my small network of colleagues at other schools if they knew any teachers who might serve as good models. After receiving a number of recommendations, I used a few of my personal days to travel and observe several teachers from other districts as well as my own. I would spend the day with these teachers as they taught, taking copious notes as I watched them engage with their students. I also made sure I observed teachers of content other than science—"Good teaching is good teaching," I would say.

I distinctly remember one particular teacher, a mathematics teacher named Mary McNulty who was also her high school's math department chair. On the day I was observing, Mary was teaching Advanced Placement (AP) Calculus—pretty heady stuff, yet I kept my mantra of "good teaching is good teaching" in mind as I observed. What I observed really resonated with me. As I walked into Mary's class, I recalled my own high school calculus experience, which was very dry and rote. I entered with low expectations, and those could not have been more wrong. Mary was making AP Calculus fun! She came into the class with a plate of "math cookies" that she would provide as an incentive for students who completed challenging problems. In the middle of the class, when she was teaching about derivatives, she spontaneously broke into a cheer: "E to the X, D-Y, D-X. E to the X, D-X!" Before I knew it, the entire class was cheering along. Talk about engaged students! These AP students were laughing and working on problems like it was a social gathering, but they were learning at a very high level! I was blown away by Mary's seamless weaving in and out of teaching calculus principles and interjecting humor. She said to me with a grin, "I try to get my students to have fun, and then I trick them into learning. They never see it coming." I came away feeling validated by seeing how her students responded and thinking about my use of humor in my own classroom.

During these observations, I noticed patterns that set these talented teachers apart from other teachers. Teachers like Mary and others I observed employed a student-focused style of teaching. Other teachers from their school would often stop by to ask questions of these teachers. Like my dad, Mary and teachers like her led by example. Other teachers recognized there was something positive going on in her classroom, as they were often tipped off by students who came from her class excited to share what had happened. I now had an understanding of teacher leadership. I noticed that teacher leaders do not self-identify but are singled out by peers who recognize their talent. If you want to find the teacher leaders in a building, their colleagues will point them out.

Leadership Requires Courage

As I continued my career as a teacher and department chair, as a science policy person in both state-level and nonprofit organizations, and finally as a professional development provider, I have been privileged to know many wonderfully talented teachers and administrators. Although they all have their own unique style in how they do their jobs, they are certainly more alike than different. For instance, they are all courageous. Leadership requires courage, which Ernest Hemingway is said to have described as "grace under pressure" in a letter to F. Scott Fitzgerald in 1926. Within the scope of education, I define courage as an individual's choice and willingness to address uncertainty. As teachers, we are dedicated to our students and make every effort to provide the best learning experiences we can. The following discussion offers an example of how courage plays out in science education leadership.

Speaking as a collective, we are constantly looking for ways to improve how we help our students learn. Research and the subsequent professional development reinforcing these new findings are constantly being developed and shared with teachers. To say that it can be overwhelming for teachers to sift through these new approaches and put them into practice would be a gross understatement. Take, for example, *A Framework for K–12 Science Education* (NRC 2012), the guiding document that inspired the *Next Generation Science Standards* (*NGSS*; NGSS Lead States 2013). Since 2013, 45 states have either adopted or adapted the *NGSS* as their state science education standards, bringing about a sea change in how science is taught and learned. The *Framework* envisions a three-dimensional approach to education in science and engineering in which student thinking is structured by their application of crosscutting concepts while actively engaged in science and engineering practices to deepen their understanding of the core ideas in these fields. Few science educators would argue against the importance of this vision. Putting the innovations of the *Framework* into practice by implementing three-dimensional science teaching and learning is another story. Implementation of this breadth and magnitude clearly requires courage.

As is true when implementing any new practice into instruction, teachers must "unlearn" what they have been taught through preservice and in-service training. As a result, many K–12 teachers of science go into this process in a somewhat timid manner. Changing the way that they have been teaching throughout their career is scary. In my experience, educators are not quick to "experiment" with their students by placing their faith in something new. Teachers must clearly connect how these new teaching strategies or techniques will make a positive difference in how they teach and how their students learn science in their classroom. For teachers to place their trust in something new, they have to be confident in the process to put it into practice. In other words, they have to summon their courage.

I have had the occasion to work with many teachers over the years, providing professional development on how to bring three-dimensional science education standards into practice. I often sense hesitation and concerns on the part of teachers as they collaboratively work through the session. I know they are wrestling with uncertainty and the fear that if they try to bring this new strategy into their classroom, it may not work. To counter this fear and uncertainty, I ask teachers to recall an experience when they tried something new. I ask them to remember how uncomfortable they were when they let their students go and explore, for example. Teachers often share how the students learned differently and were really excited about the investigation. Some teachers share how the students bring forward outcomes that the teacher was not aware of and how excited the student was to "teach the teacher" something new. By relating how they summoned the courage to try something new, the teachers realize that the risk of failure was far outweighed by the potential for success.

It has been my experience that teacher leaders do this all the time. They constantly look for new ways for students to use their natural curiosity to learn more deeply about science. The decisions that a teacher leader makes to "try something new" are calculated and well thought out. They realize there is always a risk that the lesson might "bomb," but the potential for the student to have a positive experience drives the decision. These teachers realize that even if the lesson may not hit the mark, they still learned from the experience and can apply what they learn to future lessons. This is a sign of leadership.

Leadership Requires Learning and Innovation

All leaders must keep learning and growing throughout their lives and careers. They acknowledge that leadership is maximized through struggle and failure. Think of an accomplishment that you are most proud of. Whether this accomplishment is training and completing a marathon, gaining an advanced degree, or losing 20 pounds, it does not happen unless a person devotes himself or herself to achieving the goal. This process is not easy—something worthwhile seldom is—but the effort is rewarded in the end. Teacher leaders connect their courage with innovation. They choose not to take the safe route, but instead challenge themselves to try something new in anticipation of greater outcomes. I use the expression "try it on Monday" often during my professional development sessions to encourage participants to translate their experience in professional development into professional learning in their classroom, measuring their progress by applying new ideas to their own knowledge and craft.

Teachers become paralyzed by fear of failure too often. They are bound by taking the safe way, such as using a time-honored lesson that has always worked, regardless of whether or not the lesson may be aligned to the curriculum or standards. Although this may seem like the way to go, it often leads to stagnation in creativity and innovation. I often say that if you do the same thing over and over again, there is no positive growth. Quite the contrary, you are more likely to see a decline when you use a more conservative approach to your craft. Creativity and innovation are the byproducts of a courageous approach to teaching and learning. Teacher leaders strive to learn new approaches and techniques to bring into their practice, whether through professional development, course work, or collaboration with other teachers. As a result, they become the pioneers who blaze the trail for other teachers to follow. Leadership requires making well-thought-out decisions in anticipation of amplifying science teaching and learning.

Leadership Requires Perseverance

Another defining feature of leadership is perseverance. Perseverance means sticking to a plan. Leaders always have a plan, whether it is for an hourlong meeting or the implementation of a five-year strategic plan. Lesson plans, curricula, and anticipated student outcomes are all examples of the ways in which educators use planning. The difference that is exemplified by the teacher leader is that teacher leaders are fastidious in their adherence to whatever plan they are following and consistently look for feedback that the plan is working. Leaders look for success metrics along the way, analyze these metrics often, and adjust accordingly. Similar to the way in which formative assessment is used in the classroom to monitor progress and make adjustments to instruction, leaders are constantly analyzing and adjusting the pathways they have set forward to attain desired outcomes. Positive outcomes come to fruition over time, especially if a plan is followed, studied for progress, and adjusted when necessary. A famous quote from Winston Churchill sums up the approach to perseverance quite simply. During the Battle of Britain in World War II, in a speech at the Harrow School on October 29, 1941 (America's National Churchill Museum n.d.), Churchill buoyed the British citizenry by saying, "Never give in, never give in, never, never, never, never—in nothing, great or small, large or petty—never give in except to convictions of honor and good sense." My colleagues in education do not have to lead a nation through a war, but our work is arguably as important when placed in the context of educating our children. The 18th-century British essayist Samuel Johnson might have said it best: "Great works are performed not by strength but by perseverance."

Conclusion

I feel honored and fortunate to count myself among the tens of thousands of science educators in the United States. My own career, and particularly my experience in education, has been shaped through collaborations with leaders at every level within education. As I hoped to make clear in this profile, leadership in education—especially by teacher leaders—is seldom a goal that a person sets out to accomplish. Leadership is not a quality but rather a tool that can be applied to achieve a specific purpose, such as making a difference in science education. As is the case with any tool, proficiency can be increased over time with use.

Become a keen observer and learn from people you feel are effective leaders. Remember the quote from Charlie McLaren: "You can learn as much from a bad manager as you can from a good manager." Harvest the good qualities of people you see as leaders, and learn from those who may be less effective.

Be courageous. Take stock of your inner courage, be brave, and embrace change. Challenge yourself to improve your practice in a thoughtful, measured, and collaborative manner. Do not simply run blindly into something new, but do your due diligence. Look at the research and talk with colleagues to make decisions that have the potential to result in the best outcomes for you and your students. Collaborate often and ask for feedback on your ideas. Teacher leaders are effective listeners and questioners. Sharing a new idea or strategy often becomes the start of your use of leadership as a tool.

Persevere. Always have a plan and stick to it. Whether it is long-term or short-term, for your own professional learning or for your students' learning, persevere and analyze the various metrics of progress.

Finally, don't be someone who you are not. Go with your strength. Take an inventory of your qualities and embrace those that make you who you are. Remember that working hard does not mean you have to be miserable. Find the joy in what you do, and others will gravitate toward your joy. If you do, your leadership will encourage our young people to pursue their curiosity and, in turn, change the world.

RODGER'S INSIGHTS AND INTERPRETATIONS

Peter's first lessons on leadership came from his father. These lessons included the need to develop and display a work ethic, use a sense of humor, and treat others with respect, and all stayed with Peter throughout his career, whether working in the supermarket, teaching middle school students, or providing leadership for the larger science education community.

In his profile, Peter returned to and elaborated on several important themes: Use humor, model person-centered relationships, lead by example, and keep learning and growing. I was impressed with Peter's clear and compelling emphasis on courage and perseverance and what these qualities encapsulated, such as overcoming a fear of failure as an example of courage and executing a plan as an example of perseverance.

CHAPTER

17 Learning to Lead Curriculum Implementation

JAMES B. SHORT

James B. Short is a program director at the Carnegie Corporation of New York in New York City.

NOTE FROM RODGER

James B. Short is the program director for Leadership and Teaching to Advance Learning at the Carnegie Corporation of New York. His work in philanthropy focuses on supporting teachers and school leaders and developing high-quality instructional materials for implementing new standards. Before this work, Jim was a director in the education department at the American Museum of Natural History in New York City and the science curriculum coordinator for Denver (Colorado) Public Schools. His classroom experience includes teaching secondary school science. He earned a bachelor's degree in biology from Rhodes College; a master's degree from Vanderbilt University's Peabody College, and a doctorate in education from Columbia University's Teachers College.

This profile offers many details of Jim's leadership and interest in the role of professional development as an integral aspect of curriculum reform based on high-quality instructional materials.

Reflecting on Leadership

"It is our choices that show what we truly are, far more than our abilities."
—Albus Dumbledore (Rowling 1999, p. 333)

I found writing about my experiences with leadership to be challenging. Although I knew the opportunities during my career had built on each other, writing about them in ways that might be helpful for others was new. Writing has always been more of an academic endeavor, and this request was asking for something more personal. As I reflected, I began to see that Albus Dumbledore was right—the choices I have made have taught me to stay the course. Change takes time and leadership requires patience. Many of my choices have centered on curriculum reform in science education, which continues to be an important touchstone for me. Having a vision and a plan to lead curriculum implementation is critical, but learning how to help lead others to embrace the vision and develop ownership of the plan is even more important. One of my early lessons in learning to lead came from my father.

An Early Lesson in Leadership

I grew up in Franklin, Tennessee, a small town about 14 miles south of Nashville. My grandparents had a dairy farm that had been in our family for four generations. My dad grew up on the farm and worked at the local bank. He was an elder in our church and involved in other civic activities. He also was the leader of a Boy Scout troop in our town. One Saturday morning when I was 9 or 10 years old, Dad took me to the home of a boy in town he thought might be interested in Boy Scouts. The boy lived in a poor section of town and his mom was a single parent. Dad spent a couple of hours on the front porch working with the boy to learn the Scout oath and laws, which were requirements to earn your first rank. At the time, I did not understand why we were spending so much time on a Saturday morning at this boy's house. A couple years later, when I was old enough to join the troop, I realized Dad knew that if the boy could earn the rank of Scout, he might choose to learn more about scouting and meet other boys his age.

I learned an important feature of leadership from this experience: helping folks figure out how to get started. My father was trying to help the young boy begin a journey of learning by choosing to join Scouts. Leaders in education need to consider how they help teachers get started with using new instructional strategies or choosing a new curriculum. I continue to revisit the idea of choices we make throughout these reflections on leadership.

Role of Curriculum in Learning to Teach

When I began teaching, I thought a good teacher explained things well, which is what I had wanted in a teacher. The teacher who was mentoring me had recently chosen a new biology book called *BSCS Biology: An Ecological Approach*. I did not realize it at the time, but learning to teach biology with inquiry-based instructional materials would have a major impact on my development as a science educator. I thought I was prepared to teach with an undergraduate degree in biology from Rhodes College and a master's in science education from Vanderbilt University, but I soon found out that six years of higher education were not enough when it came to teaching students who were not interested in science. Over time, I learned how to teach specific topics in biology and engage students in inquiry-based learning by choosing to trust high-quality instructional materials and what they could teach me about how students learn biology.

Leaders in science education face numerous challenges, such as aligning existing curriculum materials with new science standards. Learning to use high-quality, inquiry-based instructional materials was pivotal in my development as a new science teacher. Many science teachers today are expected to design curriculum using phenomena with science and engineering practices and crosscutting concepts. How can teachers design learning experiences using approaches they haven't experienced as learners themselves? Teachers deserve access to the highest-quality curriculum that has been designed for new standards, and leaders play a role in providing this access. There is a long-standing myth that creative lesson planning is the mark of a great teacher. A more consistent, equitable, and commonsense approach would be to relieve teachers of curriculum development responsibilities and let them focus their energy where it matters most for student outcomes: on classroom instruction.

CHAPTER 17

Role of Curriculum in Developing a Vision of Learning

After two years of teaching high school biology at a boys' boarding school, I chose to move to a K–12 girls' school as the new science department chair. The head of school wanted to introduce teachers to a new elementary science curriculum. At the time, I did not know much about elementary school science, but again I trusted that if we chose high-quality instructional materials, we could use them as a foundation to implement a new approach in the school for how students learn science. BSCS (Biological Sciences Curriculum Study) had just released an elementary science program that was designed using an innovative constructivist approach to learning science called the BSCS 5E Instructional Model. In addition to helping elementary teachers in the school learn to teach science, the 5E Model provided a concrete approach to science instruction to help other teachers reflect on their teaching practices.

My vision of learning science through inquiry-based teaching was heavily influenced by the BSCS 5E Instructional Model. This model provided a research-based approach to guide instruction that helped students take ownership of their learning. After working with elementary and middle school science teachers using the five *E*s, I now had instructional materials designed for how I wanted to teach biology.

It was during this time that the National Research Council first released new standards for science education. I had moved on from the girls' school and was now teaching and leading the science department at a co-ed preK–12 school in Brooklyn, New York. Working with the 5E Model and using new science standards challenged us to reevaluate what students needed to know and be able to do in K–12 science education. Scientific inquiry was more than a way of teaching—it was considered a content domain alongside the life, Earth, and physical sciences. The abilities to do, and the understandings about, scientific inquiry were part of this expanded standards-based approach. This perspective on inquiry-based learning and teaching meant students needed opportunities to do science themselves as well as understand the work of scientists. How could instructional materials help guide these different learning experiences for students? What types of professional development experiences were needed to help teachers learn about these ideas?

During this time, I learned that part of leadership is building buy-in and bringing others along. This can be hard to do as a young leader when you are still trying to figure out your own vision and plan. With only two years of teaching experience at a boys' high school, I was not fully prepared to take on the challenge of leading the science department at a K–12 girls' school. I had to learn how to develop a shared vision for teaching science and figure out what teachers needed to develop ownership of that vision. During my experience as a teacher and department chair, I learned to use national reports and high-quality instructional materials to inform the development of a coherent science program and a compelling vision for learning science. I also was beginning to learn more about the importance of professional learning.

Role of Curriculum in Professional Learning

Questions about the intersection between learning how to use new instructional materials and the best ways to engage teachers in professional development were at the forefront as I left the classroom with 10 years of teaching experience. I joined the staff at BSCS to lead a project supported by the National Science Foundation (NSF) to help secondary teachers select inquiry-based instructional materials and learn to use them effectively. Working with colleagues at the K–12 Alliance at WestEd and the project staff at BSCS, we developed a three-year professional development program, the National Academy for Curriculum Leadership (NACL). The NACL was designed for district-based leadership teams focused on selecting and implementing standards-based science instructional materials. To provide teachers with a manageable process and applicable set of tools, we used the Analyzing Instructional Materials (AIM) process and tools to support curriculum implementation. AIM focused on having teachers ask questions, gather information, and choose instructional materials based on evidence. AIM encouraged teachers to think about the importance of instructional materials in the learning process for students and to the instructional process for teachers.

The project also provided me an opportunity to conduct a study for my dissertation. The purpose of my research was to identify the conditions for supporting professional development that involves standards-based reform and the use of instructional materials that were considered high quality at that time. Based on the findings from the study, there were several broader implications for professional development designed to support curriculum implementation. One implication was that choosing instructional materials could be a professional development strategy using an evidence-based process. Teachers benefited from developing a conceptual flow of the science content from a unit of instruction. The process of identifying the "big ideas" in each activity and how these concepts were linked to each other showing concept development throughout a unit was critical to both choosing new instructional materials and learning how to teach with them. Teachers also learned that choosing high-quality instructional materials could help them translate ideas in the standards into instructional practices that guided student learning.

Another implication of the study highlighted the importance of monitoring the change process. Curriculum implementation is complicated and takes time. Much like their students, teachers benefit from multiple opportunities to explore new instructional materials. Teachers in the study experienced "layers of change," or learning during the selection process, in their teaching and within their district. The implication for professional learning was that leaders need to be aware of these "layers of change" that occur when choosing new instructional materials and learning how to implement them. Leaders also needed to be prepared to monitor and adjust their strategies based on the concerns of teachers.

The implications for leaders seem clear. Curriculum matters, but how teachers use curriculum matters even more. As schools and districts continue to make shifts in the instructional materials they use, teachers deserve the highest-quality professional learning to support curriculum implementation. Leaders in science education must create the necessary conditions at the school and system levels for curriculum-based professional learning.

Change management is one necessary condition. Successful change management starts with the knowledge that change isn't something that just happens—it's a process, not an event. And this change begins at the individual level, is deeply personal, and requires learning. Learning something new always precedes change; otherwise, what is the motivation to try something new?

Curriculum-based professional learning is rooted in a vision of teachers as learners. Much like their students, teachers benefit from multiple opportunities to explore new instructional materials. But they also are under constant pressure to ensure their students succeed, even as they replace trusted practices and familiar materials. The stages of change and cycles of learning suggest that leaders consider ongoing support that provides a necessary foundation for lasting improvement.

It is also important for leaders to stay the course. In the midst of change, adults often look to restore their equilibrium, including by backsliding into old ways of thinking and doing. Ongoing reflection and feedback can help teachers remain connected to the process of change. These experiences are rooted in instructional coaching and facilitate curriculum-based professional learning experiences that challenge teachers to review evidence on student learning and re-examine their beliefs about the role of high-quality instructional materials.

Curriculum Implementation in a District

After five years at BSCS, I moved to Denver (Colorado) Public Schools as a science curriculum coordinator. NACL now had a new meaning. Instead of offering technical assistance to other school districts, I was challenged to apply these experiences to the district where I worked. We used the AIM process and tools to choose new middle and high school science curricula. A group of teacher leaders field-tested the curriculum over the course of a school year. Their work informed districtwide implementation the following year. Within three years, we were rolling out new science curricula in elementary, middle, and high school classrooms in Denver. It was the hardest job I have had in education.

As we continued our implementation efforts, some of the teachers who had field-tested new instructional materials became facilitators of professional learning. To support these teacher leaders, we chose a specific approach to designing professional learning called the Science Immersion Model for Professional Learning (SIMPL), in which teachers experienced the curriculum materials as learners in professional learning. They put on a student hat as participants in lessons. Led by a facilitator, teachers became student learners and temporarily set aside their subject-matter expertise. Following this immersive experience, teachers returned to their usual vantage points to reflect on their experiences as students and revisited their initial assumptions about inquiry-based instruction and the new curriculum. It was a simple concept: Teachers needed to experience the sort of inquiry-based learning we expected them to provide for their students using the instructional materials as a key part of their learning.

Leaders responsible for curriculum implementation focus their time on professional learning. These experiences are shaped by learning designs that build on teachers' empathy for their students and challenge their beliefs about what students can do. In doing this work, leaders learn how to shift teachers' perspectives. Rather than tell teachers about curriculum, let them experience it for themselves. Putting on a student hat can help teachers trust that student-led discussions can be productive and anticipate questions and ideas that will likely surface. Leaders learn that teachers benefit from engaging in the same sort of inquiry and reflection in which students will engage. Leaders also learn that professional learning needs to extend past the launch of a new curriculum. Veteran teachers continue to implement the learning designs introduced in orientation sessions and support each other in preparing to teach future units and lessons.

Designing Tools, Curriculum, and Professional Learning for the NGSS

By the time *A Framework for K–12 Science Education* and the *Next Generation Science Standards* (*NGSS*) were published, I was back in New York City as the director of a center for science teaching and learning at the American Museum of Natural History (AMNH). With the country now focused on a new vision for science learning and teaching involving three-dimensional learning and investigating phenomena, we were excited to dive in to using science and engineering practices and crosscutting concepts with core ideas in science disciplines and engineering. Over the next few years, our center at the museum collaborated with colleagues at BSCS and WestEd to develop and field-test new resources called the Five Tools and Processes for Translating the *NGSS*. Building on what we had learned using AIM, we developed a set of tools to help teachers translate the *NGSS* into instruction and assessment. This task turned out to be much more challenging than we initially expected.

For another project, we partnered with curriculum experts at the Lawrence Hall of Science to develop an *NGSS*-designed middle school science unit on disruptions in ecosystems using the 5E Instructional Model. We chose to use SIMPL to develop a curriculum-based teacher professional development model that researchers at the University of Connecticut studied as part of the NSF-funded project. Part of this work included developing specific tools to help teachers work with students in a scaffolded way to construct scientific explanations and arguments based on evidence and reasoning. The research component of the project found that shifts in teaching practices required extensive opportunities for teachers to learn about, rehearse, and practice relevant teaching strategies. The instructional approach included extensive use of classroom discourse and student talk, which were found to be critical in helping students construct explanations and argue from evidence.

The SIMPL approach to designing and facilitating professional learning also was chosen by the Urban Advantage (UA) science initiative in New York City. The center at AMNH led this work involving teams of science educators and lead teachers across eight informal science institutions in New York City, working with more than 800 middle school science teachers in nearly half of the middle schools across the five boroughs. The focus of the program was on helping students conduct science investigations that included controlled experiments, field studies, engineering design experiments, and secondary research using large data sets. Teachers in the program conducted their own investigations while learning to use the resources at UA partner institutions. At AMNH, we collaborated with scientists at the Cary Institute of Ecosystem Studies to develop a set of river ecology data resources to investigate a local phenomenon, the effects of zebra mussels on the Hudson River ecosystem. These resources included data visualizations based on more than 20 years of data and videos of scientists at work, which later informed and became part of the Disruptions in Ecosystems unit designed for the *NGSS*. This project provided an excellent example of how to teach the abilities and understandings about scientific inquiry described in earlier science standards, as well as engage in the science and engineering practices in contemporary science standards described in the *Framework* and the *NGSS*.

We often associate leadership with a decisive nature and a singular approach to accountability. But leadership looks different when leaders model and guide inquiry rather than simply transmit expertise. They share responsibility and decision-making to promote a common purpose and collective responsibility for student outcomes. Leadership transcends titles and happens in teams, with individuals who have different skills and expertise working toward shared goals. Leaders' work resembles teaching in an inquiry-based classroom. Like their students, teachers learn from their experiences, construct new understandings, and build on what they know about curriculum and instruction. Similarly, school and instructional leaders don't pretend to have all of the answers; instead, they ask questions, actively listen, and test new ideas. Leaders model vulnerability and resilience, which encourages teachers to take risks, test assumptions, and break down old ways of thinking and doing. In other words, leaders are learners too. As a leader for each of these projects that involved designing tools, curriculum, and professional learning for the *NGSS*, I needed to put myself in the shoes of learners, teachers, and professional learning facilitators to ensure that what we designed could work for them, not just the development team that created the tools.

Open Education Resources and Curriculum

After nine years at AMNH working with New York City teachers and school leaders, I accepted an opportunity to join the education program at Carnegie Corporation of New York, the foundation established by Andrew Carnegie in 1911. Making the transition from leading projects to supporting others to do them through grants was a big shift. Now I was supporting others to do the work. The Carnegie Corporation had played a critical role in standards reform, and I had a lot to learn about grant making and philanthropy.

Carnegie Corporation's investments in science education included being the primary funder for the development of the *Framework* and the *NGSS*. Carnegie also funded several organizations to support states' adoption of new science standards. As most states adopted new science standards, the foundation's priorities began to shift to implementation, but the development of new instructional materials designed for the *NGSS* was lagging. Many teachers, schools, and systems chose to create their own materials and programs so they could begin implementing the new standards. Although there was growing demand for new science curricula, the supply was practically nonexistent. As a new program officer leading a portfolio that included our investments in science education, I knew we needed to choose a path forward. What was our theory of action to support the implementation of new science standards? What was a scalable approach for addressing the supply and demand for new curricula and aligned professional learning?

To help address these questions, Carnegie provided support to analyze and develop a report on the landscape of the *NGSS* implementation. We provided support for BSCS to revise the AIM process and tools to better align with the *NGSS*, which resulted in a new set of tools and professional learning process called NextGen TIME (Toolkit for Instructional Materials Evaluation). We learned—through Achieve's use of the EQuIP (Educators Evaluating the Quality of Instructional Products) rubric for science—that most teacher-made lessons and units did not align with the *NGSS*. Just as my work in Denver and New York City had shown, translating standards into instructional materials could help teachers keep their focus on teaching students rather than searching for resources or developing curriculum on their own. We knew from teachers using the Five Tools just how difficult it was to help teachers develop *NGSS*-designed curricula themselves. We also knew that teachers were spending several hours each week searching for instructional resources online and several more hours creating their own materials. This was not a scalable approach to curriculum reform and implementation.

We decided to launch a new initiative to improve the supply of, and demand for, high-quality instructional materials based on the *NGSS*. The goal of OpenSciEd was to ensure any science teacher, anywhere, could access and download free, high-quality, locally adaptable full-course instructional materials that support equitable science learning. Other foundations joined us, and we named the initiative Open-SciEd to build on the idea of education resources that are released under an open license that permits no-cost access, use, adaptation, and redistribution by others with no or limited restrictions. If teachers are going to search for free resources online, we wanted to make well-designed science curricula a top search result that those teachers would consider worthy of their time. Just as my dad tried to help that young boy choose to learn about Scouts, we could help teachers choose to learn about high-quality instructional materials designed for the *NGSS*.

From the beginning, the OpenSciEd curriculum development process was designed as a collaboration that brought together different perspectives and expertise, including state-level education leaders, education researchers, instructional materials developers, teachers, students, and other experts in science and education. The initiative also would design curriculum-based professional learning materials that focused on having teachers experience the same kind of inquiry-based learning we expect them to provide for their students and make the professional learning materials available with an open license for use and adaptation.

Over the course of four years, 10 states partnered with OpenSciEd and recruited more than 100 school districts and 200 middle school science teachers to field-test the curriculum and provide feedback on how to improve it and build demand in their states. Field-test data were collected from nearly 6,000 students who experienced OpenSciEd units and informed the development process. External reviews of all the OpenSciEd units using EQuIP were conducted twice during the development process by Next-GenScience (originally housed at Achieve and now at WestEd) to ensure all units received a high-quality review before they were publicly available. During this time, more than 25,000 educators registered with OpenSciEd and downloaded more than 300,000 items from the OpenSciEd website. Several of the original OpenSciEd partner states began to spread the implementation of the curriculum in their states by offering professional learning sessions to support district adoptions. In addition, more than a dozen professional learning organizations became interested in becoming certified providers of OpenSciEd's innovative approach to curriculum-based professional learning.

OpenSciEd has demonstrated how high-quality instructional materials aligned with the *Framework* and the *NGSS* can be designed and impact classroom practice. Given the inclusive and responsive development process, OpenSciEd also demonstrated how teachers can build demand while simultaneously improving the supply of high-quality *NGSS*-designed curricula. OpenSciEd's model has helped transfer ownership of the designed instructional materials to teachers as they learned to use new curriculum as intended and adapt materials to meet students' learning needs.

Having arrived at Carnegie Corporation after the release of the *NGSS* and several states' adoption of new science standards, I found the focus was on shifting to implementation. As a new program officer leading this work, I found that building supply was easier than building demand. The leadership challenge was to develop a strategy to build smart supply and catalyze smart demand. Smart supply requires a more strategic and coordinated effort among stakeholders. We chose to bring together a group of experts in curriculum development, learning science research, and professional learning to work with a group of state science leaders and science teachers to develop and field-test new instructional materials for the *NGSS*. For more than four years, these experts and state science leaders have met each month in ongoing, focused convenings facilitated by OpenSciEd. As leaders, we needed to provide opportunities for regular communication and sharing to address the disaggregated nature of the current supply and better coordination and collaboration during the development of coherent high-quality instructional materials and aligned professional learning. Smart supply requires a simpler way to describe the innovations in the *NGSS* that focuses less on *what* is in the standards and more on *how* to teach in ways aligned with the standards. We also decided the curriculum would be open and online to broaden access and adaptation for local contexts. These choices also helped catalyze smart demand. Ten states, hundreds of teachers, and thousands of students are not necessary to inform the development of new curricula, but they are needed to build demand. Early in the process, state science leaders asked OpenSciEd to help them build capacity for leading their own professional learning sessions to support field-test teachers. As a result, each state now has a cadre of professional learning providers that schools and districts can utilize to support implementation efforts.

Curriculum-Based Professional Learning

I recently marked my five-year anniversary at Carnegie Corporation. In addition to our ongoing work on implementation of the *NGSS*, we have found an opportunity to engage the field of science education in the design features of curriculum-based professional learning. While the importance of high-quality instructional materials has become better understood by many in the field, the need to shift from curriculum training sessions to curriculum-based professional learning is less understood. Again, the implications are clear: Curriculum matters, but how teachers use curriculum matters more. As schools and districts continue to make shifts in the instructional materials they use amid ongoing challenges, teachers deserve the highest-quality professional learning to support curriculum implementation. Improving practice through curriculum-based professional learning requires a focus on instruction that includes both student learning and teacher learning. Improving student learning at scale involves focusing on the instructional core—raising content standards, creating new instructional materials, and supporting teachers with curriculum-based professional learning.

In collaboration with several grantees, Carnegie Corporation published a challenge paper, *The Elements: Transforming Teaching Through Curriculum-Based Professional Learning* (Short and Hirsh 2020), which explores the effects of professional learning anchored in high-quality curriculum materials. The design features of effective curriculum-based professional learning involve core design features (curriculum, transformative learning, and equity), structural design features (models, time, and collective participation), and functional design features (learning designs, beliefs, reflection and feedback, and change management). The necessary conditions for curriculum-based professional learning (leadership, resources, and coherence) define the expectations of system leaders and principals for supporting teachers (See also, Short and Hirsh, 2023). When used in combination, design features and enabling conditions allow teachers to experience instruction as their students will, change instructional practices, and lead to better student outcomes.

Curriculum-based professional learning is grounded in the premise that teachers participate in the same rich, inquiry-based learning that their students receive. By experiencing the curriculum as learners as well as teaching their students with new instructional materials, teachers gain the confidence to shift their classroom practices, which can change outcomes for students. Through repeated cycles of learning, teachers try new instructional practices, reflect on and revise old habits, and change their practices and beliefs over time. Once teachers experience this type of learning and the impact on students, they change their understanding of what effective instruction looks like and how high-quality instructional materials play a role. This approach to teacher change is the result of deliberate design and requires attention to both the structures and pedagogy of professional learning anchored in the use of high-quality instructional materials.

Final Reflections on Leadership

The connections between leadership and learning are important parts of my journey. For me, learning is about leading change. You might say change is evidence that learning has occurred and that it is about leading the exploration of new ideas, perspectives, and mindsets. This type of change is also personal, so leaders need to understand how people learn and make meaning from their experiences. From my early learning of how to become a science teacher, my experiences learning how to lead teachers in a school or across a school system, and my time supporting grantees and learning from their work, my understanding of the importance of high-quality curriculum and well-designed professional learning has continued to guide my journey.

One recommendation I have for leaders is pay attention to the curriculum. High-quality curriculum helps translate what students need to know and be able to do in ways that provide teachers support for what to teach and how to teach. Instructional materials designed to increase student learning convey a view of teaching largely as a process of provoking students to think, supporting students as they work, and guiding students along productive paths to reach the intended learning outcomes. Professional learning experiences for teachers model these instructional approaches intended for use with students by becoming the way teachers learn to implement new curriculum. Helping teachers choose and learn to use high-quality curriculum is a choice leaders make. Designing curriculum-based professional learning in ways that teachers can experience the curriculum as learners is another choice leaders make.

Another recommendation for leaders is rethink professional learning. When leaders connect learning with leading, their work begins to look different. It means making choices that help teachers learn. Powerful learning is the result of deliberate design. Strong curriculum-based professional learning can engage educators in thinking about how students learn, how coherent instructional materials support learning, how specific teaching strategies support the research on learning, and how we assess student learning to determine deep understanding of content and problem solving. This approach to professional learning is intentional, not accidental. Leaders need to be clear about their purpose and expect curriculum-based professional learning to challenge beliefs and assumptions, mirror the instructional strategies students will experience, and promote reflection.

My final recommendation for leaders is to stay focused on the instructional core. Improving student learning at scale involves raising content standards, creating new instructional materials, and supporting teachers with curriculum-based professional learning. Improving the supply of high-quality instructional materials designed for new standards is necessary for scaling up curriculum implementation, but this is not enough. Building demand for and expert use of *NGSS*-designed instructional materials can be accomplished through more coherent approaches to curriculum-based professional learning. This, in turn, will determine the quality of implementation and help leaders scale curriculum reform in science education.

Conclusion

Just as Harry Potter got advice from Dumbledore about the importance of our choices, I think the choices teachers and leaders make matter too. When we frame our choices as opportunities to learn, we help ourselves and others lean in to opportunities to change. This reminds me of what my dad taught me by helping that young boy begin a journey of learning. It is ironic that educators are experts on learning, but we do not embrace change. Learning is all about change, and leadership is all about helping others learn to change.

RODGER'S INSIGHTS AND INTERPRETATIONS

Across his career, Jim has demonstrated the wisdom of Albus Dumbledore's statement that leads this chapter. Beginning with the experience of using high-quality instructional materials, Jim made career choices as a science teacher, professional development designer and provider, curriculum coordinator, director at AMNH, and foundation program officer to support programs that integrated the implementation of instructional materials with professional learning. These choices made clear his growing interest, knowledge, and abilities as a leader in science education.

His profile exhibits many details about his leadership development and offers several key insights: Leadership requires learning. Learning leads to the realization of personal meaning. Instructional materials matter, and how they are used matters more. Curriculum-based professional learning is essential. And, finally, leaders must stay focused on the instructional core as they pursue reform in science education.

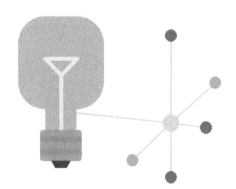

PART V

Some Leaders in
Science Education Have
Long, Varied, and Valued
Careers

CHAPTER

18. Developing Leadership in Science Education

KATHY DIRANNA

Kathy DiRanna is the director (emeritus) at the K–12 Science Alliance at WestEd in San Francisco, California.

NOTE FROM RODGER

The K–12 Alliance is a WestEd program that focuses on making school- and department-wide changes through STEM-related programs. Kathy DiRanna has shaped California's science reform efforts for the past 33 years and continues to advocate for reform efforts through the implementation of the *Next Generation Science Standards* (*NGSS*).

Nationally, Kathy collaborated with BSCS (Biological Sciences Curriculum Study) and Achieve to design the Next Gen Toolkit for Instructional Materials Evaluation (TIME). She also collaborated with the American Natural History Museum and BSCS to design 5 Tools and Processes for Translating the *NGSS* to high-quality instructional materials. Kathy served as the program coordinator for the National Science Teaching Association's (NSTA) 2006 national conference.

Kathy has a bachelor's degree in biology from California Western University, San Diego, and a master's degree in zoology from the University of California, Los Angeles. I have known and worked with Kathy for more than two decades and can attest to her long, varied, and valued leadership.

Early Years

April 16, 1947, 9 a.m. The explosion was deafening for miles. More than 600 people were killed that day in Texas City, Texas. Among them was my father, a chemical engineer for Monsanto Chemical Company. I was eight weeks old. My mother had just fed me and put me on the couch under the bay window for my morning nap. In a split moment, our lives changed forever. The bay window crashed on top of me, and my frantic mother pulled me from a pile of broken glass. She didn't find any cuts or bruises, and she reminded me as I grew up that she held me up to the heavens crying, "You were saved for something!" Thus began my journey into leadership, although I knew nothing of that at the time!

Like many of you reading this profile, I grew up in a pretty normal setting, going through the ups and downs of childhood, my teen years, and young adulthood. Being a leader wasn't a conscious goal, but as I reflect, I realize I have been considered a leader by others throughout most of my life. There were many circumstances, traits, and events that led me to this role.

After the explosion, my mother remarried and we began a new phase as a family that eventually would include four children. I was the oldest and embraced first-child syndrome: I was pleasing, diligent, reliable, conscientious, controlling, overachieving, responsible, and quick to take charge. One of my early leadership moves was to make my sister eat soap as we took baths together!

I was not cautious or afraid to do something out of the norm. I wanted to know how things worked and to explore my world. I had a very vivid imagination and knew that if we could just get our rubber band motor to turn over, our airplane made of boxes would fly! I knew there was always another way to solve a problem.

I was the only girl in the neighborhood. I learned early that to get in the game meant you needed to be serious about winning. I never had to find my voice—I was brazen enough to think my ideas mattered and I was going to share them, no matter what happened. I later learned that collaboration is much better.

My mother was a wonderful champion. She was all about justice, equity, fairness, your "good" name, and a focus on being a better person. Her daily mantra was that you could be anything you wanted to be, and you should be the best at whatever that was. Her words followed me into my professional life, and I have not lost that idea.

My mother, of course, was a major influence. She was the PTA president, Campfire Girl Troop Leader, and eventually California Teacher of the Year. She showed all of my siblings that we could make a difference. I was on the student council in junior high and high school, president of the Methodist Youth Fellowship, and dorm president and president of the Women's Association in college.

I was a Campfire Girl, so I grew up selling candy mints to strangers! In doing so, I learned the importance of setting goals, forming plans to reach those goals, and making decisions along the way. I learned the important skill of how to work with people—learning how to talk to people of all ages and work with teammates. I also learned the importance of having a quality product that could be promoted with honesty and enthusiasm. Salesmanship is a part of leading because you can help others see the potential in new ideas and new ways of thinking and doing.

Since I was a young girl, I wanted to know why: why things worked, why we did things a certain way, why only some people got what they wanted, and so on. I was taught to respect authority, but in high school I realized that those in authority did not always have the best ideas, plans, or intentions. As president of the Girls League, I realized our adviser was out of touch with what a service organization might do. She was more concerned with appearance than with the program, more concerned with her voice than her students' voices. My classmates and I were frustrated, so I took action. My student colleagues joined me, and although we were reprimanded, the program changed for the better. This experience was a turning point in my leadership and provided me with the courage to challenge processes many times in my professional life.

High school is challenging for many students. Who is in the "in crowd" (or not)? Who is in honors classes (or not)? Who is a jock (or not)? I was lucky in that I was in honors classes, good at sports, and in the in crowd . . . until one day I wasn't. Our high school had social clubs, similar to sororities and fraternities. You had to go through "rush" and be voted in by all members. If you made it, you were "kidnapped" for the celebration; if not, you got a phone call. My phone rang at 10 p.m. It's hard, even now, to remember what that felt like and what I had to face the next day as all of my friends were pledging and I was not. I learned two valuable lessons about myself and leadership: First, your value is what you determine it to be, and second, when you are in charge of groups, treat people with dignity and respect and show them they are welcome in the community. Treat people with respect and dignity, open spaces for them to grow, and nurture them as they pay it forward.

Professional Journey to Leadership

I wanted to be a doctor and to get married and have children. In 1969, it was hard to do both. I watched my male colleagues go to medical school while I went to graduate school and then became a community college biology instructor. It bothered me that my students only cared about passing the MCAT, not about the science they were learning. I wondered why, and my curiosity brought me to K–12 education and the world of professional development.

California Work

In the early 1980s, California began educational reform in all curricular areas. California had started the Curriculum Implementation Centers in English language arts, math, and science as a statewide strategy to improve education. California was also on the cutting edge with educational technology when it instituted Teacher Education and Computer Centers. The centers were regional, and I applied for a job as a science specialist, did well in the interview, and started working in December 1983. My job description was to help teachers teach science—literally, that was it! I made my job up and in doing so learned the important lesson that if you want to know your destiny, you need to create it.

I worked at the Orange County Department of Education and started meeting with colleagues about their work with teachers, made connections with colleagues offering science workshops, and interviewed a few teachers. I also ran several random workshops that were well received.

I thought we might need to offer a cohesive summer program, and we targeted elementary teachers. In my naivety, I thought science would soar in elementary schools. I knew I couldn't do the work alone, so I gathered some high school teachers to help design and deliver the institute, which was called Process and Concepts in Elementary Science. Our goal was for teachers to take little steps toward implementing science in their classrooms. The program was a huge success. We ran it a second year and were preparing for the third year when state budget cuts dismantled the program. However, our work had attracted the attention of Tom Sachse, a science director in California's Department of Education. He had some funds available and wanted to know what I would do statewide to improve science education. I flew to Sacramento for a lunch meeting and outlined on a napkin how we could start a statewide network for elementary teacher leaders. That napkin, and the bravado of the ideas expressed on it, was the starting point for my long leadership journey of reforming science education in California from 1987 to the present day.

I knew it would be impossible to create change if it had to be done teacher by teacher and recognized the need for teacher leadership to scale any reform effort. With a $250,000 investment from the California Department of Education, the California Science Implementation Network (CSIN) began in 1987. I served as director and secretary. Twenty-five staff developers (full-time teachers) with expertise in teaching science were chosen to work at the summer institute. Teachers from 50 schools from around the state attended our first one-week training. We began intuitively, believing all along that teachers are professionals who make wise choices and want to improve their teaching for the good of their students. We were developing more effective, efficient, and productive teachers. At the end of the week, the energy was palpable, engagement was high, and participants left with knowledge about change theory as well as science content and pedagogy.

As the new leaders returned to their schools, the difficult realties of creating change emerged. It became apparent that our work needed to be situated in the whole system of education and our immediate focus needed to be on the school as the unit of change. We continued our work over several years. Our slogan became "There is no *done!*" We embraced the notion that everyone had something to contribute, we could learn from each other, and no one person had *the* answers. In other words, we were a community of learners.

During the late 1990s and early 2000s, the K–12 Alliance expanded its statewide leadership role as the professional development provider of the California Mathematics and Science Partnership (MSP) grants and as the co-designer of the curriculum-based professional learning program FOSS (Full Options Science Study) Leadership Academy. The K–12 Alliance partnered with districts and universities in all of the MSP cohorts to provide professional learning in science content, pedagogy, and leadership. Through this work, we learned to include change at the district level as part of the leadership and implementation design and to strengthen our work with university faculty.

From the late 1990s through the early 2000s, California took a major step back in its progressive approach to education. The 1998 California Science Standards were an affront to all who had worked diligently to provide quality science for all learners. As a science community, we needed to speak out. I pulled on my lessons about challenging the process, selling an idea, and maintaining tenacity and the belief that every student had a right to be scientifically literate and that science should be a core subject. We attended board meetings, sat on the standards committee, rallied colleagues to speak at meetings, and tried to gain allies outside our community. Our voices were heard, but we paid a price. Some of us were eventually excluded from state committees.

But we did not go away. We continued to work with districts that had the foresight to continue inquiry-based science. And when the national movement for the *NGSS* began, California was ready. We were one of the 26 lead states, and when the state board adopted the standards in 2013, the board room erupted in cheers. Those of us who had waited for the burden of our 1998 standards to be lifted embraced the new way of thinking for science education. The K–12 Alliance, because of its expertise in science professional learning and its leadership in state policy, received funding from the S. D. Bechtel, Jr. Foundation to lead a six-year California NGSS Early Implementation Initiative.

The broad goal of the initiative was to support initial implementation of the *NGSS* in eight districts to inform decisions and set the stage for statewide implementation. The experiences of the early implementers, as well as the tools developed through the initiative, were expected to make it easier for other California districts to implement the *NGSS*. The initiative focused on changing district policies to support science education as a core subject, building leadership among administrators and teachers, changing pedagogy to align with the shifts required by the *NGSS*, and building a science learning community.

Through the work of early implementers and other K–12 Alliance programs, there was a new momentum at the state level for organizations to work together strategically to bring statewide awareness of the standards and then help districts as they implemented them.

While the K–12 Alliance has contributed greatly to statewide programs for quality science education, it has also created opportunities for leaders to join the leadership ranks wherever they are, grow in their leadership, and impact ever-expanding circles of influence throughout the state. One mark of this impact was noted as the San Diego County Office of Education was seeking a new hire for the science leadership position. Eleven out of the twelve candidates had been part of the K–12 Alliance at some point in their careers! Among our alumni are three superintendents, one college president, and hundreds of K–8 principals. Our participants have held leadership positions, including as president, in the California Association of Science Educators and other organizations. Many teacher leaders have served on state committees concerning instructional materials selection. Finally, faculty from institutions of higher education who worked with us have spurred reform in college science teaching as well.

I have been blessed to receive hundreds of e-mails, notes, and cards about participants' experiences in their leadership growth, which are summarized in these two quotes:

> *I want you to remember that you created the environment for someone like me [science phobic] to thrive and make a difference.*

> *The K–12 Alliance's leadership development and inspiration totally affected my career. During the lean years in California's science education, I informally did whatever I could to keep our district in contact with the K–12 Alliance and the cutting edge of science education. . . . I would have never been a science program manager in my district, or probably never would have stepped out of my classroom or become a principal, if it wasn't for the influence of [the Alliance] continuing to support my learning.*

National Work

Good work begets good work. This point is often illustrated by leaders who are recognized as credible, believable, and influential. Their words and presence at meetings and conferences are respected and heard. Our work has had a national presence since the early 1990s, and as a result, we have been asked by several national groups to partner in providing professional learning opportunities such as the following:

- The Center for the Assessment and Evaluation of Student Learning was a five-year National Science Foundation–funded program to improve student learning and understanding in science by focusing on effective assessment. The work resulted in the book *Assessment-Centered Teaching: A Reflective Practice* (DiRanna et al. 2008).

- WestEd's National Academy for Science and Mathematics Education Leadership was a program designed specifically to develop the knowledge, skills, and strategies needed to help education leaders become champions for effective science and mathematics education. I served as the mentor coordinator for the program, and we produced new editions of *Designing Professional Development for Teachers of Science and Mathematics* (2010).

- The National Academy for Curriculum Leadership (NACL) was developed in collaboration with WestEd (K–12 Alliance) and BSCS. I worked with Jim Short and Rodger Bybee to develop a three-year implementation plan for NACL. NACL was the crucible for people like Jim to be mentored, grow, and spread his leadership wings. Participant John Spiegel, as another example, attended as a physics teacher from San Diego and is now the director of curriculum and instruction for the San Diego County Office of Education, as well as a major contributor to California's science reform work. Another participant, Jody Bintz, became the associate director for strategic partnerships and professional learning with BSCS Science Learning.

With the national introduction of the *NGSS*, several organizations came together to develop tools and processes to help districts implement these standards, including the K–12 Alliance, BSCS Science Learning, Achieve, the American Museum of Natural History, and the Carnegie Corporation of New York. Players from the organizations knew one another from overlapping work, and new national leaders emerged from this work as well.

The collaboration among these leaders and organizations resulted in three key leadership lessons:

1. Leaders are stewards of people and programs. They see the need to give back to the system—to nurture and mentor budding leaders and provide support for experienced colleagues. Leaders develop or participate in networks to share ideas, successes, and challenges. They see the need for collaboration and recognize that there is much more to do than any one person could accomplish. On a daily basis, they live the idea that the whole is greater than the sum of its parts.

2. Leadership development is a team endeavor. Leadership is about learning and growing together, about shared decision making and challenging the system. We recognize that there is not a "mold" for leadership development. Resilient organizations are composed of leaders who embrace ambiguity and solve problems in novel ways and are willing to work with other like-minded organizations. This work requires extensive and intensive communication among colleagues, and everyone must recognize that learning is a journey taken together. We also believe that an organization can take on a life bigger than itself. When individuals have more credibility, the organization's credibility increases as well. Additionally, collaboration is fertile ground for new growth, and leadership opportunities can begin to snowball. In other words, "birds of a feather flock together" and through the "tipping point" strategy, once a strategic mass moves toward a vision, the rest of the group will follow.

3. Find a funder. Seek support from agencies, organizations, and foundations that can help you envision your dreams in various ways over multiple years.

Reflections on Leadership

I've been in education for 50 years, 40 of which involved building leadership capacity for science education. I first began thinking I needed to find the stars for the leadership work and ended up embracing Julian Weissglass's (1998) eloquent description of leadership as someone taking responsibility for something they care about. This may or may not be done on a large scale; its impact may be huge or minimal; it may be complex or deceptively simple. Whatever the case, when the action is near to your heart and soul, it has a chance of success.

In the beginning, I thought we needed the perfect program. I now know that if you build it, they will come! This is especially true if there is collaboration, incorporation of research and best practices, trust, strong relationships, and a willingness to forge new paths for new visions. I know to not stop or wait when faced with barriers, but instead to find ways to move past the barriers and reject no as an answer. It's amazing to think about what we accomplished in California by bringing good people together to have a shared vision and a commitment to see it through.

Through the years, I have found the following ideas to be foundational to building leadership:

1. Leadership is a people business. Leadership is about change—and the change process is people first, intervention second. As a network, we built on the strength of people by honoring them as human beings first, then celebrating their professionalism as teachers, and, finally, prompting their leadership by providing opportunities for them to stretch in a safe and supportive environment. We trusted in them until they could trust in themselves.

2. Leadership is more than teaching science. Being a good teacher is not the same thing as being a good leader, and working with adults is not the same as working with students. Certainly, we need teacher leaders who are knowledgeable about science content, instructional strategies, and student achievement. But it goes further. A teacher leader must facilitate change with their colleagues through their roles as presenters, facilitators, coaches, and consultants. Additionally, teacher leaders must develop expertise in organizational design, change theory, adult learning, management skills, decision making, public relations, work with a variety of stakeholders, and hand-holding.

3. Leading is about conveying hope. Having a shared vision, being persistent, having a sense of humor, and welcoming diversity enable leaders to radiate belief in possibilities. Leaders who are knowledgeable and have practical experience are credible, and credibility lends itself to confidence. Leaders with both traits move people. A leader driven by passion will succeed in expending energy to get the job done. A leader's vision must be vital, exciting, and clear and communicate the "we can" philosophy. Leaders who sense opportunity rather than danger, empower rather than control, and develop rather than maintain enable people and organizations to pursue their dreams.

4. Leading is about empowerment. Although leaders are often recognized for being in front, true leaders know how to support from behind. These leaders balance guidance with independence, vision with reality, leading with following, authority with shared decision making, and

"stardom" with partnerships. Effective leaders are committed to the community. They foster collaboration by promoting cooperative goals, building trust, strengthening others' ability by developing their confidence and competence, and providing opportunities to lead.

5. Leaders are made, they are not born. So said Vince Lombardi (Family of Vince Lombardi n.d.). We believe in a continuum model of leadership—not levels of leadership. Thus, you do not graduate to be a leader. Instead, you move along a pathway, building knowledge, skills, and understanding because you are grounded in leadership practice in your daily life. And because leadership is a continuum, everyone can be a leader in some capacity. Everyone has some sphere of influence in which they can effect change. Job-embedded leadership opportunities provide the milieu for growth over time.

6. Leaders take a systems approach. Leadership development in a vacuum has no chance of sustainability. It must be situated in real work within the system. Strategically choose the unit of change—school, district, regional, or state leadership—and the audience—teacher leaders, principals, district personnel, all or some of them. Select the trajectory for leadership, such as curriculum-based professional learning, assessment, district long-range planning, *NGSS* teaching and learning, or comprehensive school change through science. Reflect on components of the system that are needed for support, then identify where you have influence and where you need allies. If a plan doesn't work, change the plan, but not the goals!

7. The main thing is to keep the main thing, the main thing (Covey, Merrill, and Merrill 1995, p. 75). Effective leaders have a shared vision and work tirelessly to enact it. They have a plan of action that is grounded in the reality of what is happening in the field and guided by research and best practice. In keeping the main thing as their focus, leaders guard against premature clarity and instead provide time and space for understanding, collaboration, and permission to fail.

8. Politics is bigger than you think. We like to think that education is free from politics, but nothing could be further from the truth. From the election of the state superintendent of schools to local school policies, politics is part of how decisions are made. Leaders need to recognize this and strategize how to work the system to their advantage. Science leaders need to find allies (e.g., the English language arts community that can address literacy and science from their point of view) to advocate for the importance of science.

9. Backfill leadership. Whether in your organization or with your collaborators, think strategically about how to keep the leadership pipeline open, flowing with new people and new ideas. Use your networks to find opportunities for young leaders to emerge, consider mentoring roles that encourage maturing leaders to expand their horizons, and form new alliances that offer places for leadership to flourish.

New Leaders

If you are reading these profiles because you are thinking about becoming a leader, you most likely have a desire to complete tasks and accomplish goals, to have positive relationships and interactions, and to influence people and programs at some level. John Seaman Garns, president of the International New Thought Alliance, made a statement a century ago that remains true today (McGowan and Miller 2001): "Real leaders are ordinary people with extraordinary determinations." I applaud your interest and welcome you to this next step in your career. I want to share some thoughts for your consideration.

- It takes a village to do the work needed for all students to have high-quality equitable science instruction. There are multiple entry points for leadership in a variety of different settings. Find one that fits your passion, take a risk, get on the train, and grow with others who have the same "fire in their belly."

- Perfection is a no-growth position. Building a community of leaders requires being flexible and attentive to the ever-changing priority of districts and schools. Embrace Michael Fullan's (2013) approach to addressing educational issues, which includes building on what works and getting a clearer focus as to what comes next.

- Be prepared for leadership as a long-term commitment because the work is never done. Give up the idea that the pace of change will slow down. Remember that continuous improvement means you are taking a journey in a living system.

- Give up the idea that there is a perfect moment or time to lead. Recognize that where you are a year from now is a reflection of the choices you are making now. So, be brave enough every day to start conversations that matter.

- You need a community. Find knowledgeable partners to get support. Have as many teachers as possible at your school or district collaborate as a positive voice for science. Create a sense of belonging and provide support to one another for success. By creating a network, you that you are part of something bigger, which enables you to make even more connections.

- Join organizations such as NSTA and your state affiliate, as well as professional learning providers such as the K–12 Alliance, that provide and nurture leadership opportunities. When you participate in their programs, take Maya Angelou's advice: "Do the best you can until you know better. Then when you know better, do better" (Harper's Bazaar 2017).

- Share your progress. Whether you are the lone wolf at your school or part of an emerging group of leaders, spread the word about your work, the successes, and the pitfalls. Invite others to join you in sharing.

Conclusion

I have been blessed in my career to help plant leadership seeds, then watch them bloom and become beautiful science bouquets across California and nationally. It is an honor to watch these bouquets planting more seeds for generations to come. While nobody can go back and start a new beginning, anyone can start today and make a new ending.

RODGER'S INSIGHTS AND INTERPRETATIONS

Kathy's profile offers numerous examples of her leadership across a wide spectrum of contexts. She also provides numerous gems of advice to guide for current and future leaders. For example, her three key leadership learnings are extremely insightful: Leaders are stewards of people and programs, leadership development is a team effort, and leaders should find a funder.

Kathy's reflections on leadership are worth restatement, especially the initial reference to Julian Weissglass's (1998) statement about leadership involving taking responsibility for something you care about. The other insights need little interpretation: Leadership is a people business, leadership is more than teaching science, and leading requires conveying hope.

Kathy gained these insights and developed her advice over decades of work and through numerous connections to and contributions from her colleagues, many of whom are now leaders. Through all of her work, she believed in the potential of those in her programs and, by extension, the important role of science teachers as leaders.

19

Fulfilling a Legacy of Leadership in Science Education

BONNIE J. BRUNKHORST

Bonnie J. Brunkhorst is a professor emeritus of geological sciences and science education at California State University, San Bernardino, in San Bernardino, California.

NOTE FROM RODGER

Bonnie Brunkhorst began her career as a science teacher in the Lexington (Massachusetts) Public Schools, and she eventually became president of the National Science Teaching Association (NSTA). In following Bonnie's career, I realized that her leadership had historical origins, as her father was Herbert Clark Hubler, a leader in the early decades of the 20th century. Through her integrity and strength, Bonnie has extended her father's legacy of leadership in science education, especially for science teachers.

Bonnie received her bachelor's and master's degrees in geology from Boston University and her doctorate in science education, with a minor in geology, from the University of Iowa.

Bonnie is past president of NSTA and has served as a member of the National Research Council's National Committee on Science Education Standards and Assessment. She has received NSTA's highest award, the Robert H. Carleton Award for leadership in science education, as well as its Distinguished Service to Science Education Award. She was elected as a fellow of the American Association for the Advancement of Science (AAAS).

Early Years

I will begin by noting several important themes in my profile on leadership: my development of confidence as a leader, the internalization of personal values that guided my leadership, and my respect for those affected by my leadership, especially science teachers.

My personal confidence was nurtured in first and second grades at the John Dewey–based (Dewey [1916] 1966) Horace Mann laboratory school, Teachers College at Columbia University. Teachers didn't correct my left-handedness, in part because we had typewriters to write our thoughts. In wood shop, I built my own boat that floated. In cooking, I made chocolate pudding. In art, we made clay objects, then glazed and fired them. We had a fleet of bikes in the blocked-off city street to learn to ride, and I did! I learned I could do things if I wanted to.

In third grade in New Britain, Connecticut, my walk home was a challenge that I failed. I got lost. My self-confidence was diminished until I learned to use a city bus from school to my home. My confidence grew again. However, in school, my confidence took a hit when I was placed in the slow reader "Robins" because my oral reading was slower than other students', though I learned about dyslexia later in college.

My first opportunity for leadership came from an aspiration to have what my older brother had, a paper route for the *Hartford Courant*, earning rewards such as a radio. My parents didn't tell me girls were supposed to not be paper boys. So, I learned responsibility with leadership, by getting people's morning paper to them independently and collecting fees on Saturdays. I never earned a radio, but I learned what I could accomplish myself.

When I heard "Bonnie can now play the cello," I thought it was a wonderful gift from my fourth-grade music teacher to my mother, her fellow teacher. My cello provided me with social community through high school.

In fifth grade in a public school in Brookline, Massachusetts, I was a curious kid who wanted to understand how or why situations or the world worked. The experiences my parents provided in two cross-country drives became reference points for my curiosity and later for my academic aspirations. I liked my readings for English class. In orchestra and band, my music teachers valued my skills. Having confidence in my own ability to play music with others was my first sense of community, and I was assigned to the higher-ability group in seventh and eighth grades. I did okay academically. My self-confidence was building.

Community isolation limited my leadership experiences, as I was different from most of my classmates. I learned I was not among their early choices for teammates. On religious holidays, there would be only three or four of us in school. I knew what I wasn't, but I was unsure of what I was. One of the girls in school with me on those holidays invited me to go with her to the local church, which had a junior choir. The choir director welcomed me warmly. I could read music and had an alto voice. I could participate, and what I could do was valued. I had an out-of-school community, and my self-confidence continued developing.

In high school, official leadership expectations emerged. Band and orchestra were my school community. I did well in college-prep courses. Most classmates had other afterschool activities, so I socialized with the few kids who wandered over to the church. The church denomination had statewide gatherings that expanded my community of people like me.

The youth minister asked me to share my perspectives as a youth representative at their national convocation in Boston. This was the start of my official leadership experiences. My perspectives were valued by people I respected. The statewide youth director for our denomination asked me to represent our denomination at another large event. I met interesting, caring, and respectful adults and people my age. I was elected president of the Massachusetts United Christian Youth Movement (UCYM), the first time I was chosen for leadership by my peers. That year, President Dwight D. Eisenhower convened his White House Conference on Children and Youth, which I attended with my peers from other states. I was introduced to the National Council of Churches (NCC) Youth Department staff, including civil rights leader Andrew J. Young. My confidence grew as my perspectives were valued.

Young and Al Cox, from NCC's Broadcasting and Film Division, asked me to be the youth representative on the NCC/CBS TV collaboration about the network's youth programing. I was also elected by my peers to serve as the national UCYM secretary. I enjoyed interacting with and learning from other people about their lives and ideas.

When I began college at Boston University (BU), I thought science was interesting. The chair of the geology department, Dr. C. Wroe Wolfe, welcomed me, and I did well. The sciences and other subjects in the BU College of Liberal Arts provided a broad and interesting education.

Career Beginnings

I had my first experience with leadership in teaching when Dr. Wolfe asked me to help a student who needed geology to graduate but was struggling. I helped the student and she graduated. Dr. Wolfe and Dr. Mohamed Gheith invited me into the department's master's in geology program with a role as a teaching assistant. This led to a major career decision for me because I was also in line for a leadership position as a youth associate with funded international travel responsibilities to the World Council of Churches. I chose the master's degree opportunity, though I'm not sure why. I learned to teach geology labs and lead the required field trips through Boston's complicated geology, and I found I liked sharing what I knew with students.

I married a doctoral student in the geology department and needed to support him as he finished his degree, so I responded to an ad for an Earth sciences teaching position in Lexington (Massachusetts) Public Schools. In the interview, the assistant superintendent said, "You remind me of my daughter." My gender was an advantage! I was hired. I had to develop an Earth science curriculum for the district. As I taught the classes, I focused on making Earth science interesting and valuing the students themselves.

I took 10 years away from teaching science to raise my children. Being responsible for my three children required a great deal of learned leadership, and I had both successes and failures. In the interim, I was asked to teach some required geology and astronomy courses at several colleges, including Orange County (New York) Community College and The College of Saint Rose in Albany. The nuns and students were caring and helped me with protocols I didn't understand. I also taught geology on the University of Wisconsin–Superior's Grand Canyon rafting trip, and Paul Tyson, the geology chair, was a kind and encouraging mentor. He wrote one of the recommendation letters for my return to Lexington to teach Earth sciences. The associate superintendent remembered his assessment of me and welcomed me back. I taught Earth Science Curriculum Project, a new program with considerable innovations.

My first overt experience with gender discrimination occurred when a reassignment of classes fell to me. The men who also taught ESCP had modest geology backgrounds, but I was the one moved despite having bachelor's and master's degrees in geology. Why? The principal told me, "You're flexible." I considered filing a grievance but chose to cooperate and started teaching physical sciences to eighth graders alongside a veteran introductory physical science (IPS) teacher named Royal Doughty. My reassignment to teaching physical sciences led to our collaboration and then to our federal leadership as official energy educators with the Solar Energy Research Institute in Colorado. We looked for ways to make IPS more interesting to eighth graders and to share our classroom ventures at NSTA conferences.

Ertle Thompson, a professor of science education at the University of Virginia, presided at two of our NSTA presentations. He became an advocate and encouraged us when our labs made some teachers uncomfortable. We were welcomed by NSTA staff when a middle and junior high division was forming, and we were asked to run for positions on the board. Royal declined but told me he would share lesson plans if I applied. I was elected middle school director with leadership on the NSTA board. My thoughts about connecting with middle school members were approved at the summer Board of Directors meeting but seemed counter to management procedures.

Two National Science Foundation (NSF) staff encouraged and helped Royal and me apply for a grant to lead energy education workshops. We submitted the grant through Boston University and got two federal grants, one from NSF and one from the U.S. Department of Energy the following year. Father James Skehan at Boston College and Dr. Mohamed Gheith in Boston University's geology department were strong mentors. We held the workshops in our classroom labs in Lexington. A significant surprise occurred when a fellow teacher in our school who had moved into administration assumed an authoritative role toward us. I promised myself to always include my colleagues as equal professionals.

Robert Yager had treated me as a colleague at NSTA meetings and offered me a fellowship to start my doctorate at the University of Iowa. The Lexington Public Schools offered me a one-year academic sabbatical. I accepted both the sabbatical and the fellowship. Dr. Yager assigned his NSF Chautauqua professional development leadership project to me, and I led the project with respect for the teacher participants as equals. They integrated explorations from the project into their own classrooms and had a cheering community of Chautauqua supporters behind them.

I finished my PhD and was invited to join the faculty at California State University, San Bernardino (CSUSB). I still owed Lexington Public Schools my salary for the sabbatical and a commitment to return, but the Lexington superintendent extended my leave. CSUSB's vice president, Dr. J. C. Robinson, was happy to have new faculty in science education to help with the growing number of teachers the university was educating. Dr. Robinson mentored me as new faculty.

Experiences as President of NSTA

I was encouraged to run for NSTA president, and by then I had taught long enough and had ideas about how the organization could continue to "stand on the shoulders" of the many NSTA leaders in the past and develop new ways to serve teachers. Dr. Robinson was enthusiastic when I was elected and gave me "assigned time" for NSTA leadership. Lexington Public Schools eventually offered me early retirement, and I accepted. I am a proud retired science teacher.

As NSTA president, I needed to make appointments, and I received many requests and ideas for initiatives. I had also promised to support my fellow California teachers. I asked Paul DeHart Hurd from Stanford University if I had appointed too many task forces. His advice: Plant many seeds. Some may grow. The one that grew exponentially, fortunately, was a task force on developing science education standards. Applying what I learned from the gathering of many church types in my youth, I valued shared goals for science learning as well. I invited all the national science and science education societies that might be interested in working to develop science education standards to an NSTA meeting in April 1991, just before my presidency ended on June 1. The various science societies responded to invitation and indicated their interest in cooperating. They trusted a proposal would be developed and agreed upon with mutual input from each. A difficult proposal situation emerged that required quick intervention. I asked for help, and a solution was quickly established. Consequently, NSTA's board supported my recommendation that the leadership and development of the standards should be turned over to the National Academy of Science.

The academy took over the role of developing the science education standards. I was named to the National Science Education Standards and Assessment Committee and then the Executive Editorial Committee. I advocated for the inclusion of Earth sciences for all students, along with the traditional biology, chemistry, and physics. Rodger Bybee made sure the Earth sciences were included. I cheered for the major leadership developing the *National Science Education Standards* (*NSES*; NRC 1996). Angelo Collins quietly made the *NSES* come together, and Richard Klausner, director of the National Cancer Institute, chaired national standards committees. They are my personal heroes because they and others cared about science education and gave guidance and credibility to the 1996 *National Science Education Standards*.

When I was NSTA's president-elect, I was asked by NSTA President Hans Anderson if I would assume NSTA's representation to the Council of Scientific Society Presidents (CSSP). I was comfortable interacting with the science leaders and able to advocate for CSSP membership for science education organizations and the Association for Science Teacher Education as well. I encouraged CSSP's members to support the developing science education standards, and they passed a CSSP position statement to do so.

The California science education standards development included leadership within the California Science Teachers Association (CSTA). Delaine Eastin, California's superintendent of public instruction, asked California science teachers to develop state science education standards. Kathy DiRanna at West Ed used a collegial approach to lead the teachers' efforts. Standards were coming along well; I helped with the Earth sciences. The state legislature established a funded commission for science standards development, and a call for proposals went out. At a CSTA meeting, I offered to put the proposal through the grants office at my university. Our proposal was robust, with recognized leaders in science and science education and a board of directors that included nationally recognized scientists. My university required me to be the responsible principal investigator (PI). Our proposal had the highest review, and I was notified we had the contract. The next day, however, we were discredited without evidence in a *Los Angeles Times* article, and the commission in charge of awarding the contract decided that a late entry was viable. They decided to hire the PIs from both proposals. The political pressure was strong, as there was an effort to discredit me.

From my perspective, our team solutions to these proposals should be presented at all meetings and hearings throughout the state. We should speak up and defend well-established, strong science education and help prepare student-centered standards. My requirement for our teams was this: "If I turn around, you're there!" They were. Kathy led the development of our standards for me to present.

Our opportunity to work around the opposition involved laboratory standards, which Helen Quinn from Stanford University helped develop. We could give teachers a place to focus their student-centered teaching. Dr. Quinn's team quietly rearranged and restated grade-level lab standards in the draft and we added student inquiry as well. The laboratory standards went unnoticed and passed.

CSTA, led by teachers and assisted by higher education faculty, worked to help teachers find ways to teach student-centered science. Over time, they kept up communication with the state's changing political leaders. Science teacher leaders who served on our standards team moved into leadership positions and helped California carefully move forward. In 2013, the state adopted the *Next Generation Science Standards* (NGSS Lead States 2013).

Reflections on Leadership

My personal values, developed over time, guided my leadership experiences. I began life quietly meeting my personal challenges and goals. When asked for my participation and perspectives by people I respected, I developed confidence that I could do what was valued. When my peers asked for my leadership for something I valued, I left my bashfulness behind. Retrospectively, I can see that my leadership perspectives were based on values I developed from life experiences. I found I care about and respect my peers and students. I care about how Earth sciences have contributed to our lives. I found working together and developing shared community to be essential to achieving goals. Generally, my experience with leadership can be summarized with the following thoughts:

- Recognize opportunities for leadership. I have learned that leadership can start with similar interests. Sharing ideas for mutual goals can be interesting, and you can consider ways that you and others can lead.

- Listen to and learn from mentors. Mentors, respected leaders, and peers with similar values can encourage leadership efforts. Accepting the help of those trusted supporters was foundational to my leadership, and their encouragement has provided a safe place to fail and try again.

- Build on successful efforts from the past. Once you are in a leadership position, it helps to identify what efforts need to continue and what you might be able to add. It's helpful to find shared interpersonal relationships and a trusting community of fellow leaders and members, and it's just as important to delegate tasks and accept help.

- Maintain contact with the community affected by the initiatives. Maintaining community contact is important for leadership. Strengthening community engagement is essential for moving forward with new ideas. Colleagues and others have cheered, supported, facilitated and sometimes cleared the way for my leadership efforts. I knew that the larger the community of support, the more likely it was that an effort might succeed. I also saw leaders develop within and beyond the community that was involved in supporting the leader's role.

- Identify ways around, over, or through roadblocks. A mentor helped me understand that some roadblocks can be based on competition for leadership, ownership, or finances. I learned that when the roadblock remained constant, mentors could help identify alternatives. Some mentors who had the credibility and ability to help actually offered a way forward.

Reassigning the leadership beyond the roadblock was one option. Recognizing those with the potential for future leadership was essential to maintaining trust and continued cooperation.

• Show understanding and respect for those affected by your leadership, especially classroom teachers. Respect for those who may be affected by your leadership, including peers and students, is an essential part of my perspective on leadership. I believe working together and developing a shared community are essential to approaching common goals. Throughout my experiences, I became aware that for an initiative to last beyond one person's leadership, it had to be owned, adapted, and used by a growing community. An individual claiming responsibility for an effort that had subsequently grown under new leadership could handicap that effort's growth. I also learned conflict could kill an initiative. Thus, when conflict arises, alternatives are important to an initiative's continuation. Depending on other circumstances, however, a bold stance may be the way forward. It may also be necessary to embed an initiative in an organization's or state's policies.

• Provide constructive support to new leaders. As a leader steps aside, help successive leaders build on your leadership. Just cheering for their continuing efforts and assisting when asked can be helpful. Being a mentor by encouraging colleagues and "passing the baton" when possible is welcome support. I have been fortunate that my colleagues continue to provide their thoughtful leadership for science education.

RODGER'S INSIGHTS AND INTERPRETATIONS

Bonnie's leadership throughout her career was guided first by a clear set of values: She pursued what was right, what was good, and what was best for science education, especially science teachers. That leadership was supported by the courage to act personally and publicly. To do so, she needed self-confidence, which she discussed building over time.

Bonnie's guidance for leaders includes the following:

- Recognize opportunities for leadership.
- Listen to and learn from mentors.
- Build on successful efforts of the past.
- Maintain contact with the community affected by new initiatives.
- Identify ways around, over, or through roadblocks.
- Show understanding and respect for those directly affected by your leadership, especially classroom teachers.
- Provide constructive support to new leaders.

20

A Passion for Science and Teaching

ARTHUR EISENKRAFT

Arthur Eisenkraft is a distinguished professor of science education, a professor of physics, and the director of the Center of Science and Math in Context (COSMIC) at the University of Massachusetts Boston in Boston, Massachusetts.

NOTE FROM RODGER

My appreciation for Arthur and his understanding of science, especially physics, developed in the early 1990s when he joined the working group on content standards for the *National Science Education Standards* (NRC 1996). At the time, Arthur taught physics and was the science coordinator for Bedford (New York) Public Schools.

Arthur has a bachelor's degree and a master's degree in physics, both from the State University of New York at Stony Brook. He earned a PhD in science education from New York University.

He is past president of the National Science Teaching Association (NSTA) and past chair of the Science Academic Advisory Committee of the College Board. Arthur served on the physical science design team for *A Framework for K–12 Science Education*, which is the foundation for *Next Generation Science Standards* (*NGSS*). He is the project director of the National Science Foundation (NSF)–supported *Active Physics* and *Active Chemistry*, which introduce high-quality project-based science to all students.

Arthur has published articles in *The Physics Teacher, American Journal of Physics, The Science Teacher, Journal of Research in Science Teaching*, and *Physics Today*. His work on Fourier optics resulted in a U.S. patent for a vision-testing system, a lab manual, and an article in *Scientific American*.

Arthur has received numerous awards recognizing his teaching and related work, including the Presidential Award for Excellence in Science Teaching and NSTA's Robert H. Carleton Award.

Pursuing My Passions

A passion for science and a passion for teaching drive my professional life. I find beauty in the world and in representations of the world through science. I find delight in discovering new ways of seeing familiar phenomena. My love of science grows stronger when I explore ways to communicate this vision of the world through teaching, research, service, and curricula.

As I pursued my two passions for almost 40 years, I saw inequities too frequently that prevented students from getting access to a quality science course, a quality science teacher, or an opportunity to discover and experience the joy of science. These injustices rob students of the opportunity to learn about science and become student scientists. Although I did not originally set out to pursue social change, fighting inequity in science education has become my third passion. Today, my passions for science, teaching, and equity all motivate my teaching, service, and scholarship.

Teaching

I love teaching! Teaching challenges and excites me. It provides experiences that shape my research. I have been teaching for 40 years, and tomorrow I get to teach again. How lucky is that?

Classroom Experience

I started teaching as a Peace Corps volunteer in a village in Nepal. Throughout my experiences in Nepal, my 28 years of teaching middle and high school students, and my current work at the University of Massachusetts Boston (UMass), I have cherished all of the time I have spent in the classroom.

Being a good teacher requires me to find the right balance between my commitment to student learning and my love of science. Some people love the science but do not care about student learning—they should not be teachers but should be bench scientists. Other people's concern is the well-being of the students, often at the expense of the students learning science—they should be in other helping professions. I consistently remind myself that physics is certainly not the most important thing in my students' lives, but physics knowledge can help them *appreciate* their lives.

Early in my career, I thought that telling students about physics and modeling learning was enough to help them achieve. I think that my excitement did support student engagement. As I reflected more on my teaching, read the research literature, and discussed teaching with others, I realized that my students could become better learners if I could listen intently to their reasoning and help them reflect on their thinking.

One pivotal moment in my teaching was when I realized that students do not *want* to give the wrong answer. I began to understand that their wrong answer was plausible to them, and it was my responsibility to listen to how they arrived at this way of looking at the science content and the world. Exploring their thoughts and logic helped me guide them in their learning. It also radically changed the quality of questions I asked in class. I no longer ask a question for which I know the answer. I no longer ask questions to which I might respond, "That's wrong." I ask questions about how they view a phenomenon, what evidence they have for their conclusion, and whether there are alternative views that would be acceptable to them. I have been fortunate to have received numerous awards recognizing my teaching, including the Presidential Award for Excellence in Science Teaching. My teaching has also

been highlighted in an Annenberg video on inquiry that is used as a professional development vehicle across the country and as an exemplar of quality teaching by the National Science Foundation.

Professional Development

My teaching approach and instructional models have shaped my curriculum work and given me the opportunity to work with teachers across the country. Extending my classroom teaching to working with adult colleagues has helped me to better understand the development of teaching knowledge. When I work with teachers, I try to ascertain their needs. Using the "concerns-based adoption model" as background, I try to understand whether they are familiar with the new approach and whether they are ready to implement or augment the approach. At a very basic level, teachers can be placed into quadrants: (1) little teaching experience and little content knowledge; (2) little teaching experience and extensive content knowledge; (3) extensive teaching experience and little content knowledge; and (4) extensive teaching experience and extensive content knowledge. The teachers in each quadrant have different professional needs. Some need more content knowledge, some need more pedagogical knowledge, and some need more pedagogical content knowledge. Determining the teachers' needs is a necessary first step in providing professional development, but it is hardly sufficient on its own.

Teaching at UMass

My teaching at UMass has included science methods, curriculum design, and teaching practicum in curriculum and instruction. I also created a course called Revolution in Physics for our undergraduate honors students, in which we explore one revolution in physics every week. I also organized and taught a conceptualized content course in physics for Boston-area teachers.

I inform my students on Day 1 of class that my goal is to have them evaluate this course as the best course they have ever taken. The teacher ratings at the end of the semester have indicated that I have met that goal with some students. The statistical ratings indicate that students rate the quality of my teaching as excellent and feel that I have provided opportunities for reflection and growth.

In addition to my formal classroom teaching, I have organized and run a series of COSMIC seminars for the faculty of the College of Science and Math. The purpose of these seminars is to help professors improve their teaching by becoming familiar with research on how people learn. I have been working with undergraduates, graduate students, and postdoctoral professionals who are leading facilitated study groups in gateway science courses at UMass and local community colleges. Together, these efforts are intended to change college teaching and the culture and values surrounding teaching.

Curriculum Development

Developing curriculum materials gives me an opportunity to provide high-quality science materials to all students and communicate my love of science and teaching. The NSF-supported *Active Physics* project was an attempt by the American Association of Physics Teachers (AAPT), the American Institute of Physics, and the American Physical Society to develop a curriculum that would increase the number of students studying physics, as historically, only 20% of high school graduates in the United States have taken a physics course.

I led all aspects of the *Active Physics* project, which continues to be a vital part of my professional life. The creation of *Active Physics* required organizing hundreds of teachers and thousands of students as we wrote, pilot-tested, rewrote, field-tested, and rewrote the curriculum again. Because of this enormous effort, we were able to create a program that has contributed to a rise in physics enrollment nationwide. Today, more than 30% of high school graduates have completed a physics course. And more important, physics is now offered to many students in urban areas in unprecedented numbers. Every ninth grader in Boston Public Schools and Louisville Public Schools—as well as in Los Angeles, San Diego, and other cities—has learned physics through the *Active Physics* curriculum. Providing access to quality inquiry-based science for the first time to so many students from underrepresented backgrounds and in urban communities is probably the most important work I have done.

Following the success of *Active Physics*, I created *Active Chemistry* with the American Institute of Chemical Engineers (AIChE) and the generous support of NSF. Once again, I experienced the pleasure of collaborating with a new group of chemists and chemistry teachers. I am proud to report that every student in Baltimore City Public Schools enrolled in a chemistry course, and each (for the first time) had a textbook and equipment for their inquiry-based science. Together, *Active Physics* and *Active Chemistry* have contributed to a rise in science enrollment nationwide and, for the first time, provided quality laboratory-based science programs to high schools in large urban areas.

Science Competitions

I have been actively involved in creating competition programs that provide an alternative way for students and teachers to appreciate the beauty of science and mitigate inequities between students. Besides inspiring students and teachers, these programs create partnerships between nonprofit organizations and corporations that enhance science education and serve as models for other corporations to include science education in their philanthropic agenda.

The Duracell NSTA Scholarship Competition is the first science competition I helped create. In this competition, students would invent and build battery-operated devices. During the 18 years in which I chaired the Duracell competition, more than 15,000 students participated.

Now in its 18th year, the Toshiba/NSTA ExploraVision is perhaps the largest science competition in the world. In the competition, co-created by Marily DeWall and myself, student teams present their vision of a technology 20 years in the future. We have had more than 250,000 K–12 students tell us about their visions for the future and discuss the breakthroughs in science and technology that will be required to make their vision a reality. In doing so, students also must recognize that all technologies have both positive and negative consequences and explain the impact of their technology on the future society.

In addition to these competitions, I have also been fortunate enough to work on the NSTA and Toyota TAPESTRY Awards program that rewards innovative teachers. Along with a panel of perhaps a dozen NSTA members, we created the guidelines for the TAPESTRY program. As chair of the physical science judging committee for the first 10 years of the competition, I was consistently overwhelmed by

the creativity and energy of science teachers who proposed novel and interesting approaches to promote excellence in their schools.

One of the highlights of my teaching career was my work with the highest-achieving students in the International Physics Olympiad (IPhO). IPhO pits the top five high school students in each country against one another as individuals try to solve three theory problems in five hours and two experimental problems in an additional five hours.

Although I have been involved in numerous activities that required leadership, I had not thought about myself as a teacher leader during most of these times.

Reflections on Leadership

As an individual overseeing teacher leadership programs, I should be able to define "teacher leadership." I'm not sure this is the case.

The great French playwright Molière made a joke in 1670 in *The Bourgeois Gentleman* contrasting verse (which he knows) and the "mysterious" prose:

> *MONSIEUR JOURDAIN: What! When I say, "Nicole, bring me my slippers, and give me my nightcap," that's prose?*
>
> *PHILOSOPHY MASTER: Yes, Sir.*
>
> *MONSIEUR JOURDAIN: By my faith! For more than forty years I have been speaking prose without knowing anything about it . . .*

This is probably more profound than it is a knee-slapper to our 21st-century sensibilities. In being asked to share my personal experiences with teacher leadership, my first reaction was, "I'm a leader? For 40 years, I've been a leader without knowing it!" Assuming I am a leader, let me track some projects (large and small) that I've been involved in and try to cull from these experiences some characteristics that may be generalizable to all leaders.

After becoming a high school physics teacher and having remained in contact with some of my college professors, I had been encouraged to write some articles and lead some initiatives. I attended meetings of AAPT and enjoyed the collegial warmth. At AAPT, there is no badge recognition that alerts you to the fact that this person won a Nobel Prize, that one is an author of a major textbook, this one is the president of the association, or I am a high school teacher. We all listen to one another. I've always felt honored to be in the company of luminaries (I'm probably a physics groupie), but the AAPT meetings helped me not be intimidated. Being the low man in the physics hierarchy yet treated well by others showed me the value of respect. I hope I've shared that respect for others as I have changed positions. Let's call this the first leadership characteristic: humility.

After 10 years in the classroom and as an active member of the AAPT community, I found out that the dates for the summer AAPT meeting were moved to early June. This would be convenient for college professors but would make it impossible for high school teachers to attend because the dates conflicted with the end of the school year for some. I wrote a note to AAPT expressing my disappointment. I also explained that if AAPT wanted to encourage high school teacher membership, this was not a good strategy. The meeting dates were changed to alternate between early June and early July, then finally moved to always being in early July. In this way, I enacted positive change. This might be a second leadership characteristic: the ability and willingness to speak up for yourself and others.

Speaking up and advocating for a position led to greater presence and more opportunities. Sometime later, I was asked (along with Ron Edge) by Jack Wilson (executive director of AAPT) to "check out" the International Physics Olympiad, a relatively new competition. I was nervous about the trip to Potorož, Yugoslavia. I had reviewed all the past Olympiad exams and the problems were tough! For example, in a freshman physics course, a student may be asked to calculate the escape velocity for a satellite leaving the solar system. Solving correctly would earn full credit. On the Olympiad exam, the same solution would be worth the first of 10 points. To earn the other 9 points, the student would have to calculate energy efficiencies if the satellite were sent like a slingshot past Jupiter. What would happen if I arrived in Potorož and they presented the new exam problem and asked the observer from the United States to solve it in front of everyone? What if I couldn't solve it? I was nervous, but I also wanted to represent AAPT and gauge the value of U.S. participation in this event. So I exhibited another leadership characteristic: the drive to overcome fear and anxiety.

The Olympiad was a great event. Students took the exams, professors graded them, and recognition was given to the top scores. Everybody participated in cultural and scientific excursions, and friendships were cultivated. Ron and I returned and wrote a glowing report urging U.S. involvement in the event, but not all the stakeholders on the U.S. side were as enthusiastic. Their arguments against involvement raised my ire. One commented that we could not be competitive. Another agreed, adding that high school teachers aren't strong enough in physics. Another commented that we should just take five students from the Bronx High School of Science and forego a national search for our best students. It all seemed so unlike my understanding of America. We do compete; we don't hide. This leads to another leadership characteristic: the ability to ignore those who tell you that you or people like you are not good enough.

To convince people that the United States should participate, I brought together a group of high school teachers at the winter meeting of AAPT. Seven or eight of us were sitting around a lounge area discussing how we could provide evidence that the United States could be competitive. I asked if anyone had a really outstanding student this year. Dave Gewanter and a few others said they did. We figured we would have these top students sit for last year's exam and see how they did. After that, though, how could we find the best of the best in the country? Will Pfeiffenberger, who helped develop the AP exam, was walking by. "Will, would you be able to share a list of all AP Physics teachers in the United States for a mailing about the Olympiad?" "Sure, Arthur," he replied. And we began to see a pathway. A few weeks later, the outstanding students' exams came in, and they were scoring 3 or 4 out of 30 possible points. Not looking too good. And then Dave's student scored 16 out of 30 with no exam preparation. That score would have been almost enough for a bronze medal. We presented the score at an American Institute of Physics board meeting, and the board gave its blessing for us to do more to have the United States participate in the Olympiad. This is another characteristic of leadership: being able to identify a network of good people and work with them.

Shortly after, Jack Wilson asked me to locate students, train them, and accompany them to the next International Physics Olympiad. It was early February and the Olympiad was scheduled for July. I went to the American Chemical Society (ACS) meeting to talk with Dave Lavallee, who was in charge of our participation in the International Chemistry Olympiad. ACS had one Olympiad under its belt, and I wanted to know how they did all that had to be done. Dave tried to discourage me from entering that year, given that we had less than 5 months. I called Jack and explained that ACS had planned for 15 months before entering. Jack's response: "Physicists are better problem-solvers than chemists. Let's keep going." A week later, I was out enjoying some cross-country skiing. As I was skiing down a hill, I knew I was going to tumble as I tried to make the upcoming turn. I related this back to Jack, explaining that planning for the Olympiad felt the same way: There is a turn up ahead and I'm not sure if I can make it. Jack responded, "Don't you love that feeling?" "No, I don't!" Another leadership characteristics must be a willingness to embrace the challenge—or at least don't let it slow you down.

We identified the top 20 students and had them attend a week of training at the University of Maryland. We decided that all 20 students would be referred to as the United States Team for the International Physics Olympiad and that the 5 who were chosen to travel to England would be representatives. This would decrease the pressure that some students might experience. We observed students treat each other respectfully and heard from students that this was the first time they were with others who also loved academics. We had to convince a student not to leave the training when he recognized that he wasn't one of the top students (a unique experience for him). We were reminded of another leadership trait: empathy.

The United States was successful in the Olympiad and the program continued. Prior to leaving for our fourth Olympiad, I had lunch with Nobel Laureate Sheldon Glashow. He asked how we had been doing at the Olympiads, and I proudly responded that we had received three silver medals the previous year. He told me, "I like gold." Leaders sometimes have to be reminded that there is always room for growth. Don't sit on your laurels.

Other projects have reminded me of some of these leadership characteristics while teaching me about other essential qualities as well. When I was asked to run the Duracell NSTA competition, my immediate reactions were a combination of thrill and inadequacy. I was excited about this leadership role and nervous that I did not know enough to take the lead. The willingness to overcome fear met another leadership trait: the recognition that I know more than I think.

"Knowing more than I think" came up again when I was asked to lead the *Active Physics* curriculum project. I brought together the six college physics professors who would each be in charge of our project-based themes. At that time, project-based learning had not been adopted by the educational community; additionally, there weren't any rubrics for assessments or a high school physics course designed for all students. I remember trying to communicate a vision for *Active Physics* without really having a clear sense of the details of the program. I did know the intended audience and what they were like. I explained that all the students who would be learning physics through this program knew how to read, but few of them enjoyed reading. One of the professors confidently boasted that he "wrote in a very engaging style." I told him that John Updike also wrote in an engaging style and these students wouldn't read his work. As a high school teacher, I knew the students and what they enjoyed and when they enjoyed school. I knew what physics would interest them and how topics would have to be presented to engage them. Although all of the physics professors knew more physics than I did, there were things I knew that would be important to the development of *Active Physics*. As writing teams, we struggled to create a cohesive program. One writer quit because he wanted to write using a template and we were still trying to *invent* the template. The most painful part of the writing was having to take some people off the team. One was a mentor who had shaped my teaching and career. In spite of his creativity and insights, he converged on a solution to defining *Active Physics* too early and was not sensitive to ways in which it could be better. And that angst of having to hurt people when you don't want to is a difficult lesson of leadership: The project is more important than any one person.

The development of *Active Physics* through the writing, micro-testing, writing, pilot-testing, writing, field-testing, and one more round of writing allowed me to work with hundreds of teachers. I was fortunate to have that network of colleagues who could lead each element of the work. Marilyn Decker took charge of the pilot-testing and field-testing, while others took charge of the workshops. No project will succeed without the help, creative effort, and hard work of so many. Once again, I was reminded of the need for a network of good people I could trust and with whom I could work.

Active Physics produced more scar tissue than I had anticipated. There were people who wanted the curriculum project to fail and others who seemed threatened by what we might accomplish. There were people in positions of authority who gossiped to others that *Active Physics* was a failed project as we kept moving forward. *Active Physics* was published and adopted by many schools and continues to be a model for project-based learning. After its initial success, I was talking with someone familiar with NSF projects who told me that only 10% of curriculum projects funded by NSF actually get published. I was shocked. I had been under the impression that we must publish *Active Physics*, that anything short of publishing, disseminating, and supporting its adoption would be a failure. I thought publication was nonnegotiable. I had been smacked around and kept going. There were times it would have been so easy to give up. But we didn't. The writing teams didn't. The field test teachers didn't. The NSF didn't. That must be a big factor in leadership: tenacity.

The opportunity to be involved in these projects and so many more has provided me with a stimulating professional career and enriched my personal life. In reviewing my list of leadership qualities, I realize that some others have been hinted at and not explicitly articulated. So I add these and hope that readers can find implicit references. Leaders should be confident, have vision, and work tirelessly and make it look effortless. Leaders should be open and risk takers. Leaders should maintain perspective and seize the opportunities that arise.

I am presently leading the Wipro Science Education Fellowship program (Wipro SEF), which encourages and supports sustainable, positive change in science education in school districts across the United States. Wipro SEF guides teachers on their path to becoming teacher leaders. Seven universities work with 35 school districts and more than 400 fellows to create science teacher leader cadres that support district initiatives and work with other teachers. As I work with teachers who are becoming teacher leaders, I've watched them work with each other. Wipro SEF is a two-year program. In the fall of the first year, the fellows compare similar lessons taught in elementary, middle, and high schools. In the spring of the first year, the fellows regroup and focus on one science and engineering practice and

CHAPTER 20

how it is articulated in lessons at one grade band. Throughout the year, in their groups of five teachers, they review each other's lessons, provide warm and cool feedback, and critique student artifacts. In the second year of the program, the fellows take on projects that support district initiatives and conduct workshops with other teachers in their districts. How do they become teacher leaders? They become leaders by recognizing that the teachers in their district are their peers. The fellows all agree that there was no "leader" during the first year of their program. They rotated the leadership role; they took turns; they supported one another; they were a team, with each person stepping up to lead when appropriate. And that's the lesson of teacher leadership they learned and apply when they work with others: Let's lead together. With that as a foundation and philosophy, all the other traits of leadership will emerge.

RODGER'S INSIGHTS AND INTERPRETATIONS

Arthur uses personal examples to illustrate a number of his reflections on leadership. His examples touch on the use of humor to illustrate his characteristic of humility, which I will note as an important personal characteristic that is conveyed to others and contributes to a theme of person-centered leadership. There is further support for the person-centered approach to leadership by his discussion of his third passion of fighting inequity.

Another suggestion regarding person-centered leadership is in Arthur's discussion about listening to students' and colleagues' reasoning and having them reflect on their reasoning. This relates to how he identifies empathy as another vital characteristic of effective leadership.

Finally, when considering person-centered leadership, I would note that Arthur's advice to not believe people who tell you and others like you that you are not good enough is key to helping you build confidence in others, address concerns, and formulate a positive conception of both yourself and others. Arthur listed a number of other characteristics of leadership, such as a willingness to overcome fear and anxiety, which takes courage and is evidence that you may know more than you think. His recommendation to speak up for yourself and others is nicely summarized as advocacy, which is also connected with the idea of embracing challenges. Finally, leaders should keep in mind that there is always room for growth, so keep learning. Be confident, have a vision, take risks, seize opportunities, work hard, and maintain your perspective.

CHAPTER

21

My Evolution as a Teacher Leader

HAROLD A. PRATT

Harold A. Pratt retired as the director of science for Jefferson County Public Schools in Golden, Colorado.

NOTE FROM RODGER

Harold Pratt and I have been colleagues for more than 40 years, including the year he was president of the National Science Teaching Association (NSTA). I first became aware of Harold's leadership when he was science director for Jefferson County (Colorado) Public Schools in the mid-1970s. At that time, he was involved with the implementation of Sputnik-era science programs. Over the years, we worked together on many projects, including Project Synthesis in the 1980s, BSCS (Biological Sciences Curriculum Study) projects in the 1980s and 1990s, the *National Science Education Standards* in the mid-1990s, and projects at the National Research Council in the late 1990s.

Harold has a bachelor's degree with majors in physics, chemistry, and mathematics from Phillips University. He also has a Master of Arts in Teaching from Brown University and pursued advanced studies at the University of Colorado, Boulder.

Many of the interactions were professional, but there were numerous opportunities for personal discussions. Throughout the years, I have witnessed Harold's leadership in science education in various contexts.

Introduction

My career was not planned or foreordained to involve leadership. What I experienced and learned as I worked to better support science teachers contains lessons worth reviewing by anyone on a leadership path.

I began my career as a high school science teacher, with roots in my love for science that goes back as far as I can remember. My experiences have included teacher, district science coordinator, and curriculum executive director. After retiring, I was an associate for the content group working on the 1996 *National Science Education Standards* at the National Research Council, which capped off other rewarding experiences, such as being president of the National Science Teaching Association (NSTA). I have had a satisfying and productive career, but the purpose of discussing my evolution and growth is to share my many learning experiences and mentors.

Early Decisions

My love of science prevailed at every key career decision. The decision to enroll in a small, liberal arts, church-supported college as a science major and not a ministerial student was the first big decision I made, though it was not an easy one. Four years later, in April of my senior year, I reached another key fork in the road when my chemistry professor (Prof) caught me in the hall of the science building and asked if I would meet with the superintendent of a nearby small school district that was looking for a science and mathematics teacher. I was graduating with triple majors in mathematics, chemistry, and physics and an unsolicited offer of a position in the research and development department at Phillips Petroleum Company, where I had worked the previous summer. If it had not been for my loyalty and respect for Prof, I would have laughed at his request—this was not what I had planned or dreamed of for years. But I complied, listened to the superintendent, and took the position when I received an offer. This was not an overt decision to become a leader, but it must have been the beginning because as I reflected on the choice, I remember my desire to teach (or lead students) in a variety of sciences instead of becoming a narrowly focused research chemist at almost twice the salary.

My buddies thought I was crazy, and to some degree I wondered about my professional sanity as well. Although I was not aware of it at the time, Prof was the first of many mentors who saw potential in me and guided me at key points in my career. These mentors made significant contributions to my career, and to this day I am still learning about and appreciative for their assistance.

My Early Public School Experience and Continuing Education

As Prof had advised, the place to begin my career was in a small, well-managed school district. Perry (Oklahoma) High School was just the place. After a summer of intense practice with teaching and education courses, I was certified to begin teaching. I was nothing special there, other than being the youngest staff member. This was a "do it yourself" year, meaning there were no other science teachers and only a part-time math teacher. I learned by doing, with reasonably receptive students, an excellent principal, and a superintendent who provided no supervision and feedback but maintained a safe, orderly, and professionally anchored school, something I have grown to appreciate immensely after experiencing schools across the country. I loved this job, but the military draft and my upcoming marriage called.

With my science degree and a provision in the Selective Service procedures, I satisfied my draft obligation by enlisting in the U.S. Army Chemical Corps for six months of active training and duty, then serving seven more years in the inactive reserves. Coming home from duty and getting married soon thereafter, I found a position as an analytical chemist at a local oil company. After two years, though, I was ready to return to the profession I loved (and another reduction in pay). I located a position as a high school physics teacher in Jefferson County (Jeffco) Public Schools, the largest district in Colorado.

Larry Watts, the district science coordinator, took a professional interest in me, often using me as a sounding board for his creative and unusual science education ideas. I listened with interest and sometimes skepticism but delighted in the conversation that never seemed to relate to my teaching assignment but certainly must have had an impact on me. During my first three years in the high school, I accepted two important extracurricular assignments from Larry. The first was facilitating and selecting the speakers for a science seminar offered to a group of talented students from the district high schools, and the second was co-teaching a group of physics students with Frank Oppenheimer, the brother of Robert Oppenheimer, the director of the Manhattan Project during World War II. Frank, who had also worked on the Manhattan Project, retired from the public eye after being "blacklisted" because of his interest in the U.S. Communist Party when he was in college. This was at the height of the fear about communism, which was accelerated by the infamous McCarthy hearings. Not only did I co-teach with Frank, but I attended two years of night classes that he taught on the content and pedagogy of the Physical Science Study Committee's (PSSC) Physics program, the first of many innovative science curriculum projects funded by the National Science Foundation (NSF). I learned physics at a level beyond undergraduate courses, understanding the concepts that explained what I had previously spent so much time meaninglessly memorizing. This experience served as a major shift that slowly and permanently influenced my view of understanding and, therefore, teaching science. I had to relearn how to learn, but sitting at Frank's side, I was in awe of his love and insight for the discipline and his ability to make his affection for the field obvious. It was perfect timing for me to find a mentor who deepened my knowledge and to help others learn physics.

This understanding of science and my hunger to learn were enhanced every summer by NSF-funded summer institutes at universities across the country. For example, conducting research at Temple University on the plasma state of matter using exploding wires gave me a deeper personal sense of fundamental research. A summer at Richmond College studying the chemical bond approach (CBA), another innovative NSF curriculum project, elevated my knowledge of physical chemistry, which was useful as I taught the subject the following academic year. Although immersed in PSSC and then CBA as a learner, I began to see the power of the instructional strategies at the heart of these curriculum projects that were developed by scientists.

Being immersed in these projects helped me develop an appreciation for the professional development that accompanied the new and innovative NSF-funded curriculum projects. I needed support and guidance to use the new evidence-based instructional materials in which phenomena and investigations were used to inductively develop students' understanding of scientific concepts. These instructional systems required both the students and me to reason our way into fundamental concepts; we weren't introduced to them through formulas and authoritative pronouncement. We needed guidance and professional development to do this evidenced-based, logical thinking.

This was the beginning of a strong, lifelong career commitment to my big three aims in science education: having a deep understanding of the subject matter, creating well-developed and well-articulated curriculum materials, and providing professional development to support the implementation of these materials.

With three years of teaching under my belt, I applied for and was accepted at an NSF Academic Year Institute at Brown University, from which I graduated a year and a half later with a master's degree in teaching and incredible research and learning experiences. I was drawn to Brown because of its strong reputation for quality teaching as well as research. I experienced more good chemistry and mathematics under the guidance of L. B. Clapp, one of the founders of NSF's CBA project. The highlight of the program was more fundamental research in physical chemistry and a yearlong course on the role of science in society taught by R. B. Lindsay, a pillar in the physics department. With lecture after lecture reinforced by round after round of written assignments, I developed a totally new and deeper perspective of the broad social and scientific dimensions of the scientific enterprise. The research resulted in a master's thesis and a publication in the *Journal of Chemical Physics*.

District-Level Leadership and Introduction to Developing "Inquiry Teaching"

Returning to Colorado, I settled into my previous teaching position. Six weeks later, I was offered the position of district K–12 science supervisor. My good fortune was that almost immediately, I was assigned to work with Sondra Jackson, an experienced elementary resource teacher who had a strong interest in science and assisted me with the 60 elementary schools in the district. She was experienced in working with teachers and principals and providing professional development. Her tutoring helped me learn how to plan and present professional development experiences for our teachers, interact with administrators, and develop an organized professional life. Sondra was another expert mentor who fit the exact need I had at that point in my career. I don't think I could have had a better mentor in those formative years.

Behind the scenes was Larry Watts, who now was the assistant superintendent for instruction. Larry understood the value of strong subject matter support for teachers at all levels, especially at the elementary level, and was an advocate for the innovative "inquiry-oriented" instructional materials. While I was on leave, Larry had arranged local pilot centers in Biological Sciences Curriculum Study (BSCS), a high school biology program, as well as Introductory Physical Science (IPS) and Earth Science Curriculum Project (ESCP), both courses for junior high. A local center for Elementary Science Study (ESS), a large elementary science project, was also introduced that school year. Arranging and attending the training sessions for the teachers of these projects was intensely challenging and a major professional growth opportunity for me. This was my introduction to inquiry-oriented instructional materials used at the elementary and secondary levels to complement the physics and chemistry introduced later; these programs would soon become the backbone of our district's science instruction philosophy. The secondary subject matter content was rigorous but developed cohesively and systematically based on student-conducted investigations. The elementary materials were creative and unlike anything teachers at this level had used before. The inquiry-oriented pedagogy was well defined in the instructional materials but new to virtually every teacher, demanding extensive professional development and new laboratory equipment. The logistical demands of these programs required extensive support from the central science office, a professional commitment that makes all the difference in the success of new instructional programs. I quickly learned and maintained that desire and ability to support teachers throughout my career. These pilot projects, along with PSSC Physics, soon provided the components of a greatly revised K–12 district science curriculum.

At about this time, a new mentor entered my professional and personal life. At the time, the mentoring was almost unnoticed, but it would evolve into a significant and career-long relationship. Rodger

Bybee, then a professor at Carleton College, stopped by my office for a conversation, then invited me to write a chapter for a book he was editing for NSTA (see Pratt 1984). Rodger has become one of the most well-known and respected science education leaders and prolific writers in the profession. His request for me to write a chapter was unexpected, but I will forever be grateful that I agreed to take this task. It nudged me in a small way into the science education profession beyond my local school district and began a valued personal and professional relationship that has enhanced my skills and knowledge in countless ways.

Beginning Work at the National Level

Toward the end of my first two years of directing these four pilot projects, Uri Haber-Schaim, the IPS project director, invited me to join the staff in Boston as the coordinator of the project's pilot teachers and a contributing writer to the development of instructional materials. I had never met Uri, although I had worked closely with three members of his staff as they led the professional training of our local pilot teachers. I eagerly accepted and moved to Boston. My one-year leave from the school district, thanks to a very supportive superintendent, stretched into a second year.

The experience included working directly with Uri and the small professional staff to design and create an instructional sequence of laboratory investigations and accompanying narrative. Uri introduced the science education community to the innovative text format of including both the experiments and narrative material in a continuously developing storyline in the text. The results of the investigations were not included in the text but were developed by the accumulated results of the class all doing the same experiment. The teacher's guide provided real classroom data that the teacher could refer to when using the experimental result to develop concepts and scientific principles. The textbook guided the development of the concepts but rarely applied to phenomena; that came from the data derived from the students' experimental results. The student experience was essentially parallel to what professional scientists do, giving students a true experience with the nature of science. The paradigm for the textbook was revolutionary and opened the profession to a new way to think of textbook formats. This instructional strategy and textbook format were profound and would influence my leadership in instructional material development and guide me as I co-wrote a junior high Earth science textbook.

Return to Jeffco and Intense Professional Growth

With two years of this rich national experience under my belt, I returned to Jeffco with the goal of designing a coherent K–12 science curriculum based on the research and experience derived from the NSF-funded projects. After approximately three years of implementation and extensive professional development, we developed an inquiry-oriented elementary curriculum consisting of units from three national NSF-funded projects, as well as a few locally developed units that were integrated with the district planetarium, health education, and the outdoor education experience for sixth graders. The junior high curriculum program consisted of IPS, ESCP, and a locally developed life science program. The high school program included similar NSF courses in biology, chemistry, and physics.

Providing the physical materials and laboratory equipment, professional development, formal school board approval, and persuasion of several teachers to accept the changes required considerable energy and the expansion of my knowledge and experience of leadership. While this extensive development and implementation were underway, the science department staff grew to include four teachers from district schools. Developing a team effort on virtually every project and activity required leadership that initially involved honing my skill at selecting talented help for the department. The district was filled with talent, but finding the right people required time and effort. Once team members were on board, planning, delegating, monitoring, and assessing with the team helped me develop my organizational leadership skills.

An outstanding personal professional learning experience stands out from early in the implementation of the new K–12 program. Shortly after returning to my school district position, I attended a workshop where I heard a presentation by Maurice Belanger, a Harvard professor, on Jean Piaget's ideas for how children learn and develop cognitively, and I had an epiphany. I recognized why the instructional materials we were piloting were designed the way they were and the essential value and purpose of "hands-on, minds-on" learning. I was ecstatic and immediately launched a personal and district learning endeavor to know more about how elementary and secondary students learn science and how that should influence the development of instructional materials and their use in the classroom. We added an introduction to Piaget to our professional development and engaged Maurice to conduct a two-day workshop in which he interviewed students using Piaget's protocols, coupled with his interpretation of the results. Almost overnight, learning theory and research became a passion for me, our staff, and many of our teachers as the result of a short-term but influential mentor.

Just as we were launching a major implementation plan for a greatly revised elementary science program, Jim Metzdorf, our district director of staff development (a former science teacher and member of my science team), introduced me to the leadership team of the Research and Development (R&D) Center for Teacher Education at the University of Texas at Austin. They were developing a series of instruments to measure and facilitate the change in a teacher's knowledge and behavior during the implementation of innovative instructional materials in a program called Concerns-Based Adoption Model (CBAM). The R&D staff was looking for a site to test the usefulness of their instruments in guiding the change process in a school setting. Working together was the perfect match for a collaborative partnership that lasted five years. Gene Hall and his staff taught us how to use and interpret the data from the measurement of change data and assisted us with helping teachers receiving the staff development understand the process they were experiencing. We participated in the research effort by collecting data and revamping the implementation process based on the data. Gene and his R&D colleagues were making educational research history, and we were part of it! (See, e.g., Hall and Hord, 2011 and 2015.)

We learned that effective professional development can't be "one size fits all" but that combining the experience and knowledge of teachers with innovation makes a huge difference. Furthermore, there is a way to measure these qualities using the CBAM instruments to determine how to optimally plan and design the professional learning experience. The professional growth we gained enhanced our leadership ability and influenced our work with teachers and principals for the rest of my career.

During the approximately 19 years following my return to the district, I had multiple opportunities to contribute to the national science education community and to learn and grow in my knowledge and leadership abilities, including as a member of the BSCS board of directors for five years, a member of various committees, including at NSTA, co-writing a junior high Earth science textbook, teaching science methods courses at the university level, and more. These experiences and many others solidified and expanded my knowledge of virtually all dimensions of science education, but one is worth highlighting. In 1977, I was asked to serve as a team leader for the elementary committee of Project Synthesis, an NSF research project to synthesize three recently published reports on the status of science education in the United States. My colleagues who served as team leaders for the other committees were nationally recognized leaders in the profession, including Rodger Bybee, who reentered my professional life in a significant way. I was the rookie on the leadership team, and Rodger's consistent and wise counsel gave me the support I needed to be a contributing member of the project.

A New Position and a Move to Washington

The last five years of my career in Jeffco were spent as the executive director of curriculum, a position responsible for all the instructional programs in the district. The challenges of leadership became more significant as I faced a new and exciting learning curve of supervising the directors of all instructional programs in all disciplines for the school district. These were talented educators who responded positively to my leadership expectation that their programs were clearly communicated, supported, and assessed. To evaluate their success, we developed a plan of annual self-developed goals and measurable outcomes for each program area. The directors' annual personal evaluations were extensive self-reports that we reviewed collaboratively and submitted to the assistant superintendent for instruction.

After 32 years in the same school district, it was time for me to retire, a decision prompted by a major reduction in the district budget and the elimination of the Division of Instruction. A year later, while working in a part-time position at a local university, I received the leadership role of a lifetime when I was invited to become a member of the writing team for the new *National Science Education Standards* (*NSES*) at the National Research Council (NRC). After a summer of intense writing, I was asked to join the project staff in Washington, D.C. My good fortune was made even clearer when I discovered that Rodger was one of the three team leaders of the project. I observed Rodger's skill in managing more than three dozen teachers and scientist writers, producing numerous documents, and taking our work to the oversight committee of the NRC, an elite group with high standards and expectations. The five of us on the permanent staff spent countless hours reviewing and approving committee work and a variety of management decisions under Rodger's guidance and skillful facilitation.

The new position required a move to Washington, D.C., a decision my wife and I labored over before moving ahead with the opportunity. She was still engaged in an administrative position in Jeffco, with a few remaining years until retirement. We found a way for one of us to commute frequently for time together.

Two years later, as we concluded the project and finalized the production of *NSES*, Rodger, who had returned to a position as associate director of BSCS in Colorado Springs, asked me to join the staff to lead the revision of an elementary program and bring it into alignment with the *NSES*. The

challenge now was overseeing the rewriting of the material and developing new units with limited private funding and a short timeline. Again, Rodger modeled wise leadership and support as we worked together, providing another learning and growth experience for me.

Eighteen months later, with the elementary revision project well on its way to completion, I found myself following Rodger again. He had just returned to the NRC to become the executive director of the Center for Science, Mathematics, and Engineering Education. I accepted his call to return to NRC as the director of the Division on K–12 Policy and Practice in a project to promote the adoption of the *NSES* at the state level. This new role meant learning to work with leadership and the policies and politics of state leaders in science education.

At about this time, Rodger and Susan Loucks-Horsley nominated me for NSTA's Distinguished Service to Science Education Award. NSTA's leadership granted me the award, a heartfelt and unexpected honor. A few years later, after serving as president of NSTA, Rodger nominated me for the Robert H. Carleton Award for leadership in the field of science education, NSTA's highest award, which gave me another humbling and gratifying experience.

Returning to Colorado, Presidency of NSTA, and More Opportunities

Three and a half years later, thinking I might inch toward retirement, I returned home to Colorado. It wasn't but a few months before Emma Walton, a good friend and former president of NSTA, convinced me to seek a nomination for NSTA president. I was elected and began a three-year term as president-elect, president, and past president. The previous years of experience at the local and national levels paid off as I worked closely with the executive director, the other two presidents, and a nine-member board of directors to lead an organization of 55,000 members with a $20 million annual budget.

This three-year term was almost completed when Lauren Resnick, the executive director of the Learning Research and Development Center at the University of Pittsburgh, invited me to become a disciplinary fellow in the center. The center was engaged in providing professional development for the top-level administrators of several large school districts across the nation, helping them become instructional leaders to improve academic achievement in their districts.

In less than a year, this position evolved so that I was leading the science team of a $37 million NSF grant to a consortium of the University of Pittsburgh and the University of Wisconsin–Madison for a five-year math and science improvement project in the school districts in Denver, Madison, New York City, and Los Angeles. This meant more travel and more leadership workshops across the country. Lauren's expertise that furthered my education and honed my skills during these two projects can be seen in her book, *Education and Learning to Think* (1987). She led us in conducting seminars, workshops, and deep reading to support school administrators in leading schools that foster and teach higher-order skills, and her leadership contributed to my own evolving leadership.

Seeing this project wind down, and with more than enough travel under my belt, I accepted Uri's offer to assume the presidency of Science Curriculum Incorporated (SCI) as he retired and moved back to Jerusalem. Uri had formed SCI at just the time I retired so I could join the company as a stockholder and co-author. With the change in leadership, the company moved to Denver, and I assumed the position as president. This was a new leadership role but a somewhat limited one because the five authors were also the board of directors and the only stockholders, but mediation was frequently required to bring the differences among authors to agreement during the development of a new edition of IPS and the associated business plan.

I thought the SCI presidency would be my last significant role that would carry me into full retirement. Not so—a few years later, the NRC began a two-phase program to develop new national science education standards by developing *A Framework for K–12 Science Education* (2012), then handing the framework off to Achieve to develop what became the *Next Generation Science Standards* (*NGSS*; NGSS Lead States 2013). NSTA was invited, along with several science and science education associations, to provide feedback on the drafts of the standards. Because of my knowledge of standards and the development process, NSTA asked me to facilitate the collection of feedback and summarize it before sending it to Achieve.

With almost 64 years in science education leadership behind me, retirement has arrived with time for reflections that will be personally gratifying and helpful to emerging leaders.

Reflections on My Role as a Leader

I have not spent time contemplating being a leader per se. I was conscious of my role in school improvement, always making an effort to provide better education for teachers and students. How do we work with teachers to help students succeed through high-quality curriculum materials and effective instructional practices? I viewed my professional life as a way to make a difference, to support positive, innovative change. I tried to create policies and products that would support and improve the quality of teaching by more teachers and educators than I could personally contact. As I reflect on those challenges, tasks, and accomplishments, I now realize the theme of leadership runs through all of them.

I was conscious of my need to know more, learn new skills, and stay abreast of the current research. I spent a great deal of time attending seminars and workshops and reading the literature on how to implement instructional materials, understand how students learn, and apply this knowledge to our teaching strategies. The qualities of effective professional development were foremost in my mind and my own professional development for my entire career, but I rarely referred to it as leadership. I didn't think about being a better leader; I simply wanted to do a better job, to be more effective. My reflection and growth were not focused on acquiring generic leadership skills but on being an effective leader in the context of science education.

I did step back at one point to co-write *Local Leadership for Science Education Reform* (Anderson and Pratt 1995) with Ron Anderson, a professor at the University of Colorado. We drew on my experience in Jeffco and his research to describe how to go about improving the science education program in a K–12 district. We discussed what needed to be done and how to work with teachers, school administrators, and the public to design materials, develop programs, and implementing both. We did include a chapter titled "Facilitating Science Improvement: The Leadership to Get the Job Done" because I wanted to share what I had learned with other leaders.

A Few Last Words of Advice

What can be gleaned from my career experiences that could be helpful to other teacher leaders at various stages in their leadership development? Here are some thoughts:

1. Be aware of the contribution and value of working at the "policy" level. I remember that my only question when, as a classroom teacher, I was offered the position as district science supervisor with more than 80 schools to support was: Can I return to the classroom if this doesn't work? I loved teaching, but I soon recognized the expanded contribution I could make if I was successful at the policy level. So I said yes.

2. Focus on leading, not on being a leader. I wanted to be effective at working with and supporting others, but I didn't think of myself as a leader. Work at being effective, not on improving your image. I don't believe I ever thought of myself as a leader until my good friend Rodger asked me to think about it.

3. Learn, learn, learn! Be curious, open, and flexible, and look for ways to continue learning. Be aware of what is being proposed by the leaders and experts in your work.

4. Find a mentor and listen to that individual. If there is one thing that stands out in my career, it is the people who have affected my growth. I was wise enough to listen and respond to the generous and supportive professionals who influenced me at all stages of my career.

5. Avoid the many fads that seem to always be present at your professional door. Are they grounded in good science or good research-based pedagogy? What is the resource cost to implement it?

6. Be authentically enthusiastic. Be sure that you are yourself in every situation, whether talking one on one, leading a meeting, presenting to the board of education, or writing a report. Anything different will undercut your message and goals. Know what you believe and how to communicate it clearly and with authentic passion.

7. Know the trade-offs of leadership for your personal life. Most leadership roles are open-ended; there is always more to be done. Keep a tight calendar with time for yourself, family, and friends.

8. Know your science. Science is the basis for decisions you make. Know who you can turn to when checking the validity of a claim or the accuracy of instructional materials.

9. Most of all, enjoy, thrive. Your mental health and your success depend on it. If you ever feel like it is just a job to earn a living, take stock and change something.

RODGER'S INSIGHTS AND INTERPRETATIONS

There are a number of notable points in Harold's profile. First, along with others in this book, he did not plan for a career of leadership in science education. In fact, his original intention was to become a researcher in the petroleum industry. Through a series of opportunities and decisions, he evolved from a classroom teacher to a national leader. In a sense, the role of his leadership remained in the background of his career. He did not perceive himself as a leader. He wanted to work on school improvement and improve the quality of science teaching and student learning.

Second, Harold benefitted from the models of leadership provided by a number of mentors who clearly influenced his own development and leadership.

Third, early in his career, he formulated clear aims or purposes: understanding of subject matter, using well-developed and high-quality curriculum materials, and providing professional development to support teachers' implementation of those materials. I also note the role of the CBAM and its explicit recognition of the personal concerns of those ultimately responsible for implementing reforms.

Fourth, Harold's advice to young leaders is clear and succinct: Be aware of the contributions and value of working at the policy level. Find a mentor and listen to that person. Avoid the many fads that seem to always be present. Be enthusiastic and authentic. Know the trade-offs that you will make in your personal life as a result of your leadership. Know your science. And, most important, enjoy your efforts and thrive, but make a change if something doesn't feel right.

My appreciation for Harold's leadership has developed based on more than four decades of discussions. Harold has taught me and many others about the value of having the capacity to listen with acceptance and understanding, even when you do not agree with someone or something. He has displayed the courage to "tell it like it is," especially when you have bad news. He has a clear perception of reality, which is important for knowing potential outcomes. Finally, he models the importance of accepting others regardless of differences in characteristics such as class, education, religion, race, or gender. All of these qualities have helped Harold become a leader in science education for more than 60 years—whether or not that was his intention.

PART VI

Leadership by and for Science Teachers

Guides for the Journey

CHAPTER

22.

Eight Guides for Leaders

AS SCIENCE TEACHER LEADERS and those facilitating professional learning search for new instructional materials and develop their teaching knowledge, skills, and abilities, they make decisions. In making decisions about what program to select, what unit to teach, which standards to emphasize, and which activities to use, science teacher leaders are influenced by their values regarding leadership. Scientists, for example, prefer straightforward, parsimonious explanations rather than complex or convoluted ones. In this chapter, I summarize eight of the qualities identified in the profiles in this book that I believe leaders should use to guide their own leadership in science education.

Purpose

Purpose is the first quality I will discuss because leadership must be grounded by a purpose, and that purpose must be complemented by well-designed plans.

Effective leaders have a clear sense of purpose, meaning that they are clear about their aims and knowledge of what needs to be done. Having a purpose directs one's leadership. All of the leaders who contributed to this book implicitly or explicitly expressed a sense of purpose, a sense that daily experiences slowly and incrementally contribute to each student's continuous development of scientific literacy. Many science teachers and those providing professional development have an imagined world of scientifically literate citizens. They realize that such a goal is difficult to achieve, yet they still demonstrate a sense of purpose derived from this vision.

As a guide, purpose is analogous to a general direction for the journey's final destination—albeit one that may not be reached in a specific science class or adoption of a specific school science program. Thinking about what one may achieve in a science class—for example, having students develop science and engineering practices associated with inquiry—introduces the idea of vision, the concrete manifestation of purpose. If purpose describes a general direction for travel, vision identifies a specific destination for the journey. As an example, a science teacher's purpose involves developing scientific literacy to "prepare students for life," and the vision is that students will develop the ability to "construct scientific explanations using logic and evidence."

Most of this book's contributors identified, in some way, the need for leaders to express and clarify a sense of purpose and give justification for pursuing their stated vision (see Figure 22.1).

Figure 22.1. Statements of Purpose

Huff: Create a new vision for science education and strategies.

Rounds: Look at the whole picture. Find the *why* in what you do.

Johnston: Provide personal meaning with a *why* and *how*.

Garcia: Leadership requires a clear vision and strategic plan.

Spiegel: Ultimately, leadership is about stepping in, making a contribution, and serving others.

Wu Moriarty: Teacher leadership requires a clear vision and plan.

Kastel: Before anything, set goals and articulate a common vision.

McLaren: Leadership is a tool that can be applied to accomplish a purpose.

Short: Having a vision and a plan to lead curriculum reform is critical.

DiRanna: Leadership is about conveying hope and having a consistent vision and a plan of action.

Brunkhorst (Bonnie): Be aware of the personal values that guide your leadership.

Eisenkraft: The project is more important than any one person.

Pratt: Be aware of the contribution and value of working at different levels (e.g., developing policies).

One lesson I have learned is that once a leader has a vision, he or she must also plan for changes. If an individual only expresses a vision, there may be excitement, but the implied changes will flounder and likely fail based on questions of "What will change?" and "What does this mean for my students and me?" On the other hand, having only a general plan, such as stating a need to select new instructional materials, will lead to questions such as "How does this change relate to other ideas and proposals at the national, state, or district levels?" The effective leader must have answers to questions about the larger purpose and more specific plans.

Leading in a standards-based context means establishing a vision; translating that vision for students, professional learning committees, organizations, and schools; and maintaining the vision while adapting it to school science programs and teaching practices. Expressions of each leader's vision must be in the context of that individual's own environment, organization, or institution. Although visions need not be complex or elaborate, they must be different from current programs and practices. Visions are not statements of the status quo; they are new and substantial, and they look to the future.

A complement to purpose is a well-designed plan. Following any initial question—"Why are we changing our science education program and practices?"—the second question is "How are we going to do it?" These questions seem simple enough, but some individuals have a purpose without a plan. Others may have plans but not a vision. Providing leadership requires a plan with both directed action and flexibility. One must give a sense of direction and be responsive to those with whom one is working. Being too rigid is ineffective; being too flexible may also be ineffective. Providing leadership means continually reaffirming your vision while simultaneously adjusting your plans.

Exemplary leaders have developed a sense of design that strongly influences the composition, flow, and elements of particular plans. In general, the definition of *design* can include planning, inventing, or having a goal. I believe that as a leader, you should address the arrangement, composition, and connection among the various elements in your translation of a vision into plans. Science teacher leaders as designers continually work out relationships among elements so they can address the leader's purpose. Note that underlying this design process is a plan.

For science education leaders, the journey toward excellence begins with the formulation and clarification of a purpose. They must direct their thoughts and actions to the larger purposes of science education and imagine a world of STEM-literate citizens. Leaders must recognize that the destination defines the journey. Finally, there must be recognition that *why* one is embarking on a journey influences *how* one travels.

Relationships

Personal relationships with those affected by one's leadership are fundamental to how effective a leader one can be. Statements about the role of relationships from the profiles on leadership are presented in Figure 22.2.

Figure 22.2. Statements About Relationships

Olson: Leaders have to trust in others.

Bess: Make connections.

Rounds: Know your audience. See those who will share with you and support you.

Blake: Network with others who may be core to leadership. Recognize others' concerns.

Pullen: Leaders are caring.

Long: Be accessible, collaborate, and celebrate.

Johnston: Create a learning environment based on family (i.e., team), loyalty, and hard work.

Brunkhorst (Herbert): Develop mutual respect by valuing all individuals and identifying their strengths.

Spiegel: Recognize the value of relationships with mentors.

Wu Moriarty: Leadership is relational.

Kastel: Find opportunities for relationship building—see yourself as a leader to become one.

Pruitt: We need to proudly and boldly proclaim, "I am a teacher."

Short: Learning how to help lead others to embrace the vision and develop ownership of the plans is important.

DiRanna: Leadership is a people business and requires a community. Laugh at yourself, smile with colleagues, and pay it forward.

Brunkhorst (Bonnie): Show understanding of and respect for those affected by your leadership.

Eisenkraft: Identify a network of good people and work with them. Have empathy.

Pratt: Recognize the trade-offs of leadership in your personal life, but most of all, enjoy and thrive.

Contributors described relationships with terms such as *trust, connections, concerns, caring, accessible, support, team, mutual respect, relational, people business,* and *empathy.* These terms show that these leaders understand the need to pay attention to personal relationships. In this section, I answer the following question: What are practical and concrete guides that will help leaders with personal relationships? I propose three guides: recognize concerns about change; understand the role of perceptions of self, others, and situations; and clarify personal meaning.

Recognize Concerns About Change

Try to imagine the feelings and reactions of individuals attending a workshop on standards and hearing that your role in the standards' implementation will require you to connect crosscutting concepts, science practices, and disciplinary ideas in your curriculum; engage students in activities that have them form explanations for phenomena; include engineering design and the nature of science as complements; implement your components for a coherent progression from kindergarten through high school; and, finally, connect the activities to English language arts and mathematics.

It isn't an exaggeration to say that you would be concerned. That is the very situation that teachers faced with the announcement of new science education standards such as the *Next Generation Science Standards* (*NGSS;* NGSS Lead States 2013).

If you were leading this presentation, how would you respond to questions such as the following: What do these changes involve? Do we need to change? How will I have to change? The need to recognize personal relationships boils down to this statement: I have to change in significant ways, and I have concerns. Whether these concerns are explicitly expressed or left unspoken, leaders have to recognize them and maintain positive personal relationships.

As a guide for educational leaders, I recommend a classic resource on the issues I have mentioned, *Implementing Change: Patterns, Principles, and Potholes* by Gene Hall and Shirley Hord (2015). This book presents developmental Stages of Concern as well as Innovation Configurations and Levels of Innovation Use. The authors also describe leadership, facilitation styles, and realities of change. As you continue in the role as leader, you will find multiple reasons to refer to this book.

Understanding the Role of Perceptions of Self, Others, and Situations

This guide emphasizes a person-centered view of human behavior. For many years, Arthur Combs and others (Combs, Avila, and Parkey 1978) have investigated the characteristics of effective helping relationships, including teaching. Their research indicates that a teacher's perceptions of self, students, and the teaching task are critical to forming effective personal relationships. In more recent years, this perspective has been applied to leadership in the book *On Becoming a School Leader* (Combs, Misner, and Whitaker 1999). Effective teachers perceive other people, particularly their students, as able, friendly, worthy, intrinsically motivated, dependable, and helpful. In the same manner, effective teachers see themselves as good teachers who are needed and trustworthy and relate well to other people.

Leadership as discussed here certainly qualifies as a helping relationship. The perception of teacher leaders also plays an important role in teachers' effectiveness. Better teacher leaders see themselves assisting and facilitating rather than coercing and controlling. They identify with larger issues; are personally involved with issues, problems, and other people; and view the whole process of education as important. Effective teachers see their task as helping people rather than primarily dealing with objects. And, generally, they try to understand the perceptions and backgrounds of those they are leading.

Briefly, this view proposes that behavior is a result of an individual's perceptions of themselves, the situation, and their purposes. A state's adoption of new science standards, for example, can result in varied responses depending on the individual's view of the situation. One person may accept the new standards because they are timely and relevant, another may express concerns about the implications for reform, and still another may resist the idea of change because he or she already has a program.

Clarifying the Personal Meaning of Your Leadership

The search for meaning in science education is not a search for knowledge of science or teaching; rather, it is a personal search for answers to questions such as these: What really counts for me? What are my real concerns as a science educator and leader and as a person? The search for meaning is as old as human thought, dreams, and aspirations. This does not mean, however, that each individual can avoid asking and answering the question. Teacher leaders should pause and reflect on personal meaning as it relates to their leadership. Science teachers often encourage students to search for meaning in their experience with science, and I suggest that those leading have the courage to engage in their own search as well.

Leaders can convey their search for meaning in several ways. First, the leaders profiled in this book demonstrated meaning through their *involvement* with students, science, teaching, and leadership. Second, people *commit* to things that have meaning for them. Do exemplary teacher leaders communicate their commitment to science teaching? Yes, they do. Finally, leaders' *convictions* indicate another aspect of meaning. In every case, the profiles bear witness to the individual's commitment to students, science, teaching, and their conviction that science education can make a difference in their lives and the lives of their students. A sense of meaning guides the leaders' journeys, and they demonstrate that meaning through their involvement, commitment, and conviction.

In *When Teachers Face Themselves* (1955), Arthur T. Jersild addressed the topic of meaning, writing, "The crucial test in the search for meaning in education is the *personal implication* of what we learn and teach" (p. 80). For many individuals, this statement summarizes a personal approach to teaching and leading: They are always guided by the personal implications of their leadership. For some, Jersild's statement may ring hollow because they assume that subject matter, information, vocabulary, and facts are, in and of themselves, meaningful enough. Both research and everyday experiences suggest that some professional development does not necessarily have personal implications—that is, meaning—for participants.

Opportunities

When opportunity knocks, it is sometimes hard to know the form it will take and the courage, knowledge, and skills required to respond. Some situations motivate individuals to change out of a desire for personal improvement, while other situations impose the need to change. An example of the former is a science teacher's use of new technologies that can enhance student learning. Implementing a curriculum program because the state adopted innovative science education standards is an example of the latter cause of change.

The profiles on leadership cite a variety of perspectives related to opportunities and associated topics such as change, creativity, courage, and leadership (see Figure 22.3).

Figure 22.3. Statements Related to Opportunities

Olson: Assuming leadership takes courage.

Blake: Autonomy allows opportunities for creativity. Be willing to take risks.

Pullen: Leaders are brave.

Long: Choose your moments—quality over quantity.

Spiegel: Confidence (and courage) will resolve tensions about leadership in new opportunities.

McLaren: Be courageous and embrace change.

DiRanna: Leadership is a long-term commitment—be brave enough to begin conversations that matter.

Brunkhorst (Bonnie): Recognize opportunities for leadership.

Eisenkraft: Overcome fear and anxiety with courage; embrace the challenge, or at least let it slow you down.

Pratt: Avoid fads.

Accepting Opportunity Implies Change and Adaptation

Moving this discussion beyond opportunity, leadership involves working with and guiding peers. In the context of initiating changes, the leaders will have to recognize the need to provide colleagues with experiences that require new approaches to curriculum and instruction—that is, they will have change. This requirement presents an opportunity for leadership.

What do I mean when I say that adaptation is a quality of science teacher leadership? I use *adaptation* to mean changing to become suitable to a new or unique situation. Science teacher leaders use adaptation as a guide in two critical ways: in response to situations within their classroom and in reference to leadership beyond the classroom.

Effective leaders have a sense of adaptation that guides their interactions with peers, administrators, and the public. Adaptation consists of achieving an appropriate balance between directedness and spontaneity. Sometimes effective leadership is more directed and sometimes it has to be more spontaneous, yet leadership that is exclusively directed with no option for adaptation may be less effective. Leaders who are only spontaneous are probably on a journey without a GPS.

Summon the Courage to Lead

Individuals have choices as they realize the opportunities to lead. That realization also means they may have questions, doubts, concerns, and anxiety about making the choice to do something new and different. It takes courage to overcome concerns and make the choice to provide leadership.

Courage, as it is generally understood, is the quality of mind that enables one to face challenges with confidence, resolution, and perseverance. The courage I am describing is neither physical, such as that required in military combat, nor that discussed by philosophers; rather, it is the courage to investigate new forms of teaching practices, design curriculum programs, and develop policies, depending on the educational context and opportunities. Several contributors to this book discussed the risks they took and how they subsequently became leaders. These may seem like small changes, and in a sense they are, but all the small changes occurred when the individual overcame the anxiety associated with the opportunities and prospect of change.

Leaders need courage to continue their development and realization as individuals and professionals. The leaders in this book consistently demonstrate courage. I am not talking about heroic bravery, but everyday courage. They do not risk change and innovation once in a while; they do so every day as they lead others.

Opportunities to lead will present themselves in numerous contexts. Courage is an essential guide that opens the door to other features of leadership, such as facilitation, integrity, and equity.

Facilitation

I carefully chose the noun *facilitation* as a guide for emerging or continuing leadership. There is a truth about leadership that one must realize: A leader may call meetings, provide resources, and conduct professional development, but the responsibility for accomplishing goals and achieving changes rests on the shoulders of others. Facilitation suggests processes such as coaching, modeling, connecting, integrating, and nurturing. Facilitation does not imply controlling.

The profiles of leadership recognized, if not stated directly, many forms of my use of the term facilitation (see Figure 22.4).

What is required to facilitate a group and move it toward your purpose? Effective facilitation requires an integration of knowledge, abilities, and qualities such as understanding your purpose and goals, being flexible, managing time, actively listening, appropriately responding to needs and concerns, and providing resources.

Facilitation as a guide allows for the emergence of creative ideas and unique solutions from a team. It also recognizes the role of empowerment and the ever-present need to recognize and respond to the concerns and perspectives that emerge in the processes of change.

Figure 22.4. Perspectives on Facilitation

Bess: Make connections.

Rounds: Be bold and integrate new things.

Blake: Recognize priorities.

Long: Trust and delegate.

Garcia: Keep community participation in the process of change.

Brunkhorst (Herbert): Be an effective listener and communicator, and be a guide from the side.

Wu Moriarty: Leadership must be nurtured.

Kastel: Create new leaders. Tap into strengths and what you enjoy.

Pruitt: Model the themes you represent: presence, courage, perseverance, and equity.

Short: Leadership looks different when leaders model and guide inquiry rather than simply transmitting expertise.

DiRanna: Communicate your ideas, work, and progress—leaders must facilitate change.

Brunkhorst (Bonnie): Provide constructive support to new leaders.

Eisenkraft: Speak up for yourself and others.

Empowerment and Leadership

In *Leadership*, James MacGregor Burns (1978) points out that two essentials of power are motives and resources. If science teachers lack motives, resources diminish; if they lack resources, motivation diminishes. Science teacher leaders only do their work in relation to other people, so issues of motive and resources extend to students, administrators, and the community. This discussion of power extends to the essential idea of power as relationships with others.

Empowerment of others provides a way to make links among motives, resources, and—most important—students. The prefix *em* means "to provide with, to convey, or to put into." So, *empower* means "to provide with power, to convey power, or to enhance the power of another person."

In their book *Leaders: Strategies for Taking Charge* (1985), Warren Bennis and Burt Nanus described several dimensions of empowerment. The first dimension is developing a sense of *significance* in others. Recall the prior discussion on relationships. Effective leaders create a vision that makes others feel as though they make a difference. To be significant, this vision must have substance and transcend the superficiality of slogans. Science teacher leaders and professional developers are, for example, translating the vision of new standards into innovative standards-based science programs and sustaining the new programs by building the capacity of school personnel. In so doing, they are making a positive difference in students' science education.

A second dimension of empowerment involves developing new *knowledge, skills,* and *attitudes*. Leaders working with teachers to develop greater understanding of the science content (e.g., core ideas, science practices, and crosscutting concepts) and the skills to select, adapt, or develop high-quality instructional materials will empower individuals to implement curricular changes, leading to greater competence, a sense of mastery, and feelings of accomplishment.

Third, empowerment provides a sense of *community*. For example, when science teachers in a school system have the common purpose of implementing the new state standards, and cooperate to achieve that goal, they develop a sense of community and collegiality. Instituting professional learning communities clearly supports this dimension of empowerment.

Empowering people results in their experiencing greater *enjoyment* in their work, a fourth dimension of empowerment. Contemporary theories of motivation and leadership recognize that individuals have higher needs, including to know and understand, to engage in meaningful work, and to develop personal and professional efficacy. These needs are critical to a greater enjoyment that can facilitate leadership by and for science teachers.

Facilitating Change: Practical Guides

As leaders are called on to facilitate changes, they often encounter challenges with clarifying the changes and providing contextual examples. Facilitators have an initial guide that must be used—to clarify the changes implied by the proposed innovations. What is new and different, and what will actually change in the classroom, school, district, or state? Furthermore, who is responsible for the actual change?

The innovations from the *NGSS* present an engaging set of examples to introduce the concerns of those responsible for changes. As a facilitator of change, you should be guided by clarifying what the change is and what elements of the educational system will change. I will briefly discuss the innovations of *NGSS* as examples for this discussion (see Figure 22.5.)

On the face of it the innovations of *NGSS* present a significant and complex array of changes for leaders to clarify. The innovations are relatively easy to describe and extremely difficult to implement. Figure 22.5 summarizes implicit changes for different dimensions of the educational system.

Reading the far-right column might make the average educator feel overwhelmed by the array of complex changes. The challenge for leaders is to minimize the complexity while maintaining the integrity of the *NGSS* innovations. Some examples of clarifying and simplifying the changes might include introducing processes to select or design an integrated instruction sequence, providing examples of instructional materials from OpenSciEd, or providing background on engineering and the nature of science. The implications for leadership and professional learning also seem clear. Finally, I strongly recommend that teacher leaders review ideas about the Stages of Concern, Innovation Configurations, and Levels of Use in *Implementing Change: Patterns, Principles, and Potholes* (Hall and Hord 2015).

Figure 22.5. *NGSS* Innovations: Clarifying Changes

First-Generation Standards	*NGSS* Innovations	Implications for Changing Components of the K–12 Educational System
Single concepts in science disciplines	Integration of three dimensions (science and engineering practices, disciplinary core ideas, crosscutting concepts)	Integrated instructional sequence
Engineering/nature of science as supplemental learning	Engineering and nature of science incorporated as practices or crosscutting concepts	Context for lessons, units, and courses that includes an understanding of engineering and the nature of science
Standards as descriptions of content	Standards as performance expectations	Formative and summative assessments for explaining phenomena and designing solutions
Grade-level or course emphasis	K–12 learning progressions	Reform of school science program
Few connections to other disciplines	Explicit connections to *Common Core State Standards* for English language arts and math	Curricula, instruction, and assessments include English language arts and math as appropriate

Integrity

General definitions of integrity include terms such as *complete, whole, unimpaired, uncompromising, adherence to a set of values, responsible, quality of completeness,* and *wholeness.* Integrity comes from the Latin root *integer,* which one should recognize in reference to a whole number. Many contributors alluded to the general idea of integrity or the integrity of science in their reflections on leadership (see Figure 22.6).

Figure 22.6. Statements About Integrity

Olson: Leaders have to model resilience.

Huff: Distributed leadership includes individuals, organization, and educational agencies—we all have a role.

Bess: Dedication can lead the way.

Rounds: A good work ethic will take you to your destination.

Blake: Leadership requires courage.

Brunkhorst (Herbert): Be passionate about helping students.

Spiegel: Find passion and inspiration.

Wu Moriarty: Leadership is a practice, something an individual does independent of an assigned role. It is part of the design of effective learning organizations.

Kastel: Stay close to those your leadership affects.

Pruitt: Develop a broad and deep understanding of issues.

McLaren: Don't be someone you are not—inventory your qualities and embrace those that make you who you are.

Short: Stay focused on the instructional core.

DiRanna: Leaders must be credible.

Brunkhorst (Bonnie): Have confidence that you can do what is valued.

Eisenkraft: Have humility, tenacity, commitment, confidence, and perspective.

Pratt: Focus on leading, not being a leader—be authentically enthusiastic.

Personal Integrity and Science Education Leadership

My use of integrity as a guide for the leadership journey centers on the role of science teachers and their integrity as it pertains to the science disciplines. In *Integrity* (Carter 1996, p. 7), the author describes what he means by the term:

> *Integrity, as I will use the term, requires three steps: (1) discerning what is right and what is wrong; (2) acting on what you discerned, even at personal cost; and (3) saying openly that you are acting on your understanding of right and wrong.*

Note that this quotation addresses personal qualities of integrity as well as connections to relationships and equity. Carter's statement also applies to issues of right and wrong as they relate to the integrity of science. Carter suggests the first criterion is reflection about, for example, the value of science and the conceptual completeness of a discipline such as biology. The second criterion is that one acts on a

sense of what is right—that is, as a science teacher, does one fulfill the responsibility of maintaining the scientific perspective? Third, a person of integrity is unashamed of doing what is right.

Realizing one's personal integrity as a leader is not easy, but it is essential. Certainly, each of the three steps Carter proposes has its own difficulty. Take time to reflect on and clarify what you believe and value about science. It is too easy to avoid the personal review of what we want to convey about science to our students. Why is it important to teach science, and what is it, specifically, that ought to be the right message to leave with students? I recognize this first step is fairly easy to write yet extremely difficult to realize in practice.

The second step is not easier. It is easier *not* to act on what you believe to be right relative to teaching science or leading the improvement of a science program than it is to take action and introduce a topic that some may see as controversial (e.g., climate change, vaccinations). Leadership in the reform of school science programs likely will not be as controversial as climate change, but it will require a commitment to selecting the best instructional materials by either participating in or providing professional learning experiences that allow colleagues to gain new knowledge and skills.

In the third step, a science teacher leader is willing to say he or she is acting in the best way for students. Instead of saying, "I am *just* a teacher," it would be wonderful to hear teachers follow Stephen Pruitt's idea and report, "I have studied science and I'm a teacher, and I'm doing the best for students. That's the reason to support the implementation of the state's standards for science education."

Maintaining the Integrity of Science

I just suggested that science teachers, as leaders, have a professional obligation to uphold the integrity of science. There is a long history of efforts to change the science curriculum by introducing religious ideas and omitting evolution (Branch 2020). Proponents of this view argue that organisms are too complicated to have evolved, and they believe this complexity shows that there must have been deliberate design. This belief questions the integrity of science.

In recent years, the consequences of climate change and the COVID-19 pandemic have been added to the anti-intellectualist (Hofstadter 1963) list of topics that some use to challenge the integrity of science, despite mounting evidence from the scientific community about the harms of both.

All citizens confront science-related life situations in personal, national, and global contexts, and their understanding of the nature of scientific knowledge is limited, to say the least, showing that the scientific and education communities have failed to fulfill their roles in democratic societies. Helping citizens make scientifically reasonable and prudent decisions within personal, social, and global situations is the challenge, and it is a significant one. An enlightened citizenry would be a significant countervailing force to those perpetuating nonscientific views. The beliefs of vocal minorities such as climate change deniers and others creating controversies based on nonscientific beliefs likely will continue. My argument is that we must complement knowledge based on the science disciplines with an equal emphasis on understanding the structure and function of the larger scientific enterprise and the nature of scientific explanation. The standards include nature of science themes such as science is a way of knowing; scientific knowledge is based on empirical evidence; scientific models, laws, mechanisms, and theories explain natural phenomena; and scientific knowledge is open to revision in light of new evidence.

How can new standards, the science disciplines, and leaders in science education assure the continued recognition of expertise? One recommendation is to redouble our commitment to educating students *about* the scientific enterprise. I will end this section with a paraphrase from Adam Gamoran (2019): In our recognition and pursuit of integrity, we should let evidence light the way.

Learner

Leaders are learners. This idea is fundamental. Effective leadership requires a broad range of knowledge and skills, most of which leaders did not learn in preparation for teaching and seldom acquire in typical professional development. The different situations of leadership highlight the need to be a learner. A review of the profiles in this book provides support for the idea that leaders are learners (see Figure 22.7).

The difficulty of identifying the knowledge, values, and skills that leaders should develop is clarified by considering a continuum with unique situations at one end and common situations at the other end. For science teacher leaders, a unique situation might involve being asked to discuss a career in scientific research with a colleague who sees you as a mentor. An example of a common situation could be conducting a workshop on the implications of *NGSS* or aligned state standards for school personnel.

Figure 22.7. Statements on Leaders as Learners

Olson: Leaders should continue to expand their experiences.

Huff: Empowerment means new knowledge, skills, and attitude.

Bess: Engage in self-reflection.

Rounds: A science leader is prepared.

Blake: Listen, then respond.

Pullen: Leaders are learners.

Long: Do your homework—always be a learner.

Garcia: Listen to your mentors.

Brunkhorst (Herbert): Never stop being a learner.

Spiegel: Learn to step outside the box and ask questions such as "What if we do this?" and "Why don't we try this?"

Kastel: Stay close to other leaders to keep learning.

McLaren: Learn from all leaders you observe.

Short: Recognize the connections between leadership and learning—curriculum-based learning is rooted in a vision of teachers as learners.

DiRanna: When you know better, do better—leaders must develop expertise in organization, design, change theory, adult learning, decision-making, and how to work with a variety of stakeholders and hold people's hands.

Eisenkraft: There is always room for growth; don't sit on your hands.

Brunkhorst (Bonnie): Listen to and learn from mentors.

Pratt: Find a mentor and listen to him or her—know your science.

Reflecting on the Meaning of Content and Experiences for Those You Are Leading

The idea of personal meaning and leadership was introduced in the earlier section on relationships, but it also seems worth mentioning with regard to learning as a leader. Given the varied situations of leadership, we should reflect on these questions: How can your leadership help those you are leading in unique situations? What do you have to know, value, and be able to do to be effective? To paraphrase Jersild (1955), the main point of reflecting on the meaning of content and experiences is to identify the personal implications for what you and those affected will learn because of your leadership.

As science teacher leaders, this reflecting can provide personal answers to questions about what knowledge and skills are worth learning. As a result, leaders providing professional development for colleagues and presentations for administrators or school boards will make every effort to establish connections between the content and contexts important to those audiences.

Teacher leaders can find different ways to share meaningful content and experiences with those affected. They can involve participants in experiences that have materials that are usable and manageable within the participants' context, such as physical experiences a teacher can use in the classroom with students. They can also identify socially relevant topics that will engage participants.

Expanding Your Recognition of Different Educational Perspectives

As one's leadership expands within and across different educational situations, it can be helpful to apply a model of four different educational perspectives that may be in play. Initiatives within science education systems may represent perspectives as *purpose, policies, programs,* and *practices* for science education.

The different perspectives can be exemplified by contemporary discussions in science education. When explaining new standards and making the justification for changes in science education, individuals may reference *A Framework for K–12 Science Education: Practices, Crosscutting Concepts, and Core Ideas* (NRC 2012) and describe the purposes of science education and scientific literacy as the basis for college, careers, and citizenship. These discussions contribute to a perspective I refer to as *purposes*.

Another perspective may be expressed as a blueprint, a syllabus, or plans for science education. These are, by my definition, *policies* for science education. Policies are more concrete statements that may relate to specific aspects of science education, such as instructional materials or assessments. The *NGSS* (NGSS Lead States 2013) provide a set of policies for science and may be unique to disciplines, grade levels, or specific emphases. There is an important point to note: Standards are not the curriculum; they are *plans* for a curriculum.

When one hears about new instructional materials or assessments, these are the actual *programs* based on the purposes and policies. Finally, and reasonably so, teachers ask about methods, strategies, and teaching. These are the *practices* of teaching science. Among the more significant complexities in translating the policies of standards to the practicality of school programs and classroom practices are the categories and contexts of innovations that must be addressed.

Explaining the Categories and Contexts of Innovations

I discuss NGSS innovations here to present examples of challenges in other state standards and reforms. Clarifying the categories and contexts of innovations will be helpful in presentations to colleagues, administrators, and the public. Figure 22.8 briefly summarizes the NGSS innovations.

Figure 22.8. NGSS Innovations

Innovation 1.	K–12 science education reflects three-dimensional learning.
Innovation 2.	The *NGSS* incorporate engineering and the nature of science.
Innovation 3.	The *NGSS* provide expectations for student performance.
Innovation 4.	*NGSS* practices, core ideas, and crosscutting concepts build coherent learning progressions from kindergarten through grade 12.
Innovation 5.	The *NGSS* provide connections to *Common Core State Standards* for English language arts and math.

The five innovations present an array of challenges to the most dedicated leader: What does it mean to teach three dimensions? How does one incorporate content that one knows little about? What is a performance expectation? What is implied by K–12 learning progressions? And finally, what are the connections to the *Common Core State Standards*? For most individuals and professional learning communities, the innovations contain a significant level of complexity.

Complexity, ambiguity, and perplexity create strong forces against effective actions to implement changes based on the *NGSS,* other related state standards, and reforms. If your leadership involves addressing innovations by identifying potential instructional materials and recommending professional development, how would you proceed? As a leader, how would you replace the complexity with simplicity, the ambiguity with clarity, and the perplexity with priority? How would you do this and maintain the integrity of the *NGSS?* As a leader, what would you have to know, and where would you look for information to help you explain the implications for change in your school and district? Figure 22.9 proposes answers to these questions, based on the work of Mike Schmoker (2018, 2019a, 2019b).

Figure 22.9. Reducing Complexity and Maintaining the Integrity of the *NGSS*

NGSS Innovations	Educational Category	Implied Context	Implementation Challenge	Potential Resources
3-dimensional (3-D) learning	Content	Teaching (science and engineering practices, disciplinary core ideas, and crosscutting concepts)	Integrated instructional sequence	BSCS 5E Model (Bybee 2015)
Incorporate engineering and nature of science	Content	Teaching additional practices and crosscutting concepts	Integrated instructional sequence	BSCS 5E Model (Bybee 2015)
Performance expectations	Learning outcomes	Assessments for 3-D learning	Alignment of assessment	A Guide to Implementing the Next Generation Science Standards, Chapter 6 (NRC 2015)
K–12 learning progressions	Curriculum	Coherent school program	Selecting, adapting, and developing K–12 school programs	A Framework for Leading Next Generation Science Standards Implementation (Stiles, Mundry, and DiRanna 2017)
Connections to Common Core State Standards	Content linking basics and science	Teaching English language arts and math	Incorporating Common Core State Standards	Next Generation Science Standards, Volume 2, Appendixes L and M (NGSS Lead States 2013)

Stating that leaders in science education are learners is both an obvious and a fundamental guide. As teacher leaders accept opportunities, the need to understand more and different aspects of education becomes evident. Leaders must understand the implications of the changes for which they are advocating and address the effects on different parts of the science education system. Beyond some fundamental suggestions about understanding standards, contributing to professional development, and reflecting on priorities (see, e.g., Rodman 2018), it is difficult to describe the unique challenges individual leaders will confront.

Equity

To bring the guide of equity closer to the topic of educational leadership, in the *National Science Education Standards* (NRC 1996, p. 2), we stated,

> The intent of the *Standards* can be expressed in a single phrase: Science standards for all students. . . . The *Standards* apply to all students, regardless of age, gender, cultural or ethnic background, disabilities, aspirations, or interests and motivation in science.

More recently, the *NGSS* embraced the equity theme as "All standards, all students" (NGSS Lead States 2013, Appendix D, Volume 2).

Figure 22.10 presents statements of equity from profiles in this book.

Figure 22.10. Statements Related to Equity

Long: Recognize and value all perspectives and roles.	**Brunkhorst (Herbert):** Effective leadership includes valuing all individuals.
Johnston: Appreciate the personal experiences and abilities of others.	**Pruitt:** The critical issue is equity.
Garcia: Keep equity in mind as a goal of education.	**Eisenkraft:** Don't believe people who tell you that you or people like you are not good enough.

While proclaiming the equality of all citizens, the disparities of equity among citizens, including students, are clear and must be addressed. The consequences of COVID-19 and climate change have exacerbated these problems and increased calls for equity. I will attempt to give some initial guidance to leaders.

Basic Needs and Equity Concerns

Individuals have basic needs that must be met, such as air, water, and food. In answering questions and giving direction to those seeking guidance on equity in science teaching, I have found Maslow's (1970) hierarchy of basic needs to offer helpful and concrete responses. Whether or not an individual's needs are satisfied influences a person's motivation, thoughts, and behaviors. These needs also have implications for educators, particularly classroom teachers and professional developers.

Physiological needs such as food, water, air, and sleep have the greatest motivational force in the hierarchy. If a person is chronically deprived of such needs, he or she would be motivated to fulfill those needs first. Other needs would have only marginal influence on the person's behavior. For example, factors such as hunger, thirst, room temperature, and need for rest all contribute to a student's motivation. Students may behave in overt ways that indicate needs at this level; teachers should try to be aware of these behaviors and respond as best as they can within the school environment.

If a person's physiological needs are met, they may have needs for safety and security that could influence motivation. This level is described by a need for order, structure, stability, dependency, security, and freedom from chaos and fear. Students perceive their environment as secure when it has regular patterns and people are consistent, persistent, and dependable. To that end, teachers should maintain a physically and psychologically safe classroom. For example, teachers should help students with minor problems, avoid situations that are threatening to students, and take care not to use humiliation or embarrassment as motivators for learning. If teachers change a routine, they should inform students of the change and try to make the transition gradually.

Belongingness is the next level in Maslow's hierarchy. As many students stayed at home in 2020 and subsequently returned to school, many reported interests based on belongingness. Frustrations associated with belongingness needs are characterized by words such as *alienation, loneliness, aloneness,* and *rejection.* A student looks to give and receive affection, have a friend, belong to a group, and find a place with people and a community. Some students may feel rejected because they do not belong to groups, so they need affection and positive regard from the teacher. Teachers can show students they are interested in teaching them and that they belong in their classroom by addressing them by name, recognizing their achievements both in and out of the classroom, talking with them about topics of common interest, and listening to their ideas.

Self-esteem, the fourth level of needs, includes motivation toward developing a stable, firmly based, positive perception of self. The need for personal esteem is divided into mastery needs and prestige needs. Mastery concerns the individual's need for achievement, competence, and confidence in carrying out tasks, assignments, and projects. Prestige, on the other hand, is more social-psychological in nature, referring to the need for others to show a respectful attitude toward the individual. Status, recognition, importance, and dignity are defining characteristics of need fulfillment at the prestige level. The equity-oriented response to this need would be arranging learning experiences so that all students will gain a sense of accomplishment in your class. Teachers can give acknowledgement and praise when appropriate, and help students with difficult problems, concepts, and choices.

The highest level in Maslow's hierarchy is self-actualization, which represents the full development of an individual's potential. Self-actualization involves creativity, knowledge, ethics, and aesthetics.

Addressing basic needs as a way to improve equity shows an understanding of how students' motivations and behaviors develop. The comments and concerns that students have expressed about the pandemic and climate change show how such needs influence their daily lives.

Making Science Meaningful and Equitable

Teachers can also look for external ways to make science content relevant and interesting for all students, such as by establishing connections between science phenomena and students' lives. To help make these connections, teachers should ensure materials and activities are physically accessible for students, interesting, and relevant to them socially.

Object, events, and materials can have meaning due to their physical proximity to a student. Some meaning can be associated with an object in a student's hand, which is the reason hands-on activities are valuable. Developing concepts based on hands-on activities also can establish some level of personal integration of the content.

To engage students' interest, teachers might use discrepant events, dramatic videos, or puzzling questions. These strategies can initiate student thinking and interest in further exploration of science topics, which may result in examinations of students' different explanations and reasoning.

My third suggestion for making connections—social relevancy—is how most think of personal meaning. Although some students establish meaningful connections with local, regional, or global issues, leaders should not assume that timely issues are meaningful for students. Still, social relevancy has tremendous potential for the personal integration of scientific concepts and processes.

The use of place-based, problem-based, or project-based approaches in the classroom complements all of these suggestions and may enhance opportunities for meaningful engagement by all students.

Most of the means of addressing issues of equity require adequate personal relations with all students and enthusiasm when working with them. I found evidence in support of this idea decades ago (Bybee 1973, 1975, 1978) and think these suggestions are especially appropriate now.

Systems

Leaders will benefit from understanding educational systems, especially the dynamics of systemic change. At some point in their profiles, most of the leaders in this book alluded to the idea of systems, though they might not have used that specific word. I considered terms such as *contexts*, *connections*, and *networks* to be references to systems. My interpretations of contributors' statements are shown in Figure 22.11.

Figure 22.11. Statements Related to Systems

Olson: Leadership occurs at many different levels, and leaders have to trust in others.

Huff: New standards imply changes in programs and practices.

Bess: Connections are a key to leadership.

Ryder: Connections are an important characteristic of leadership.

Blake: Networking is essential.

Long: Recognize and value all perspectives and roles.

Garcia: Have the community participate in the phases of change.

Kastel: Leaders need a network of support.

Brunkhorst (Herbert): The role of your administrator is to help you do your job.

Pruitt: Be aware of your vision and context.

Short: The role of leaders is to create conditions at the school and system levels for curriculum-based professional learning.

DiRanna: Politics is important.

Brunkhorst (Bonnie): Maintain contact with the community affected by the initiatives.

Eisenkraft: Identify a network.

Pratt: Be aware of the policy level.

Cultivating the ability to think in terms of systems and associated concepts and how they complement aspirations expressed as one's purpose will be useful for teaching leaders (Fullan 2010).

Systems and Science Education

A systems perspective informs a leader's role. So, what are some basic ideas about systems and systems thinking? A system is an interconnected set of elements that are coherently organized to achieve a specific purpose (adapted from Meadows 2008). One could also view a system as comprising a set of relationally arranged and interdependent components that are organized as a definable entity in a given environment to attain a specific purpose (Banathy 1992).

With these definitions in mind, one can infer or identify several key terms related to systems: *components* (or *elements*), *interconnections* (or *interdependent relationships*), and *purpose* (or *function to achieve something*). One could also think about *boundaries* that define the system. Finally, there is a dynamic quality of systems, meaning that they change to achieve a purpose. This feature introduces the idea of feedback in a system. So, general features of systems that contribute to effective leadership include boundaries, components, the flow of resources and information, and feedback (Thorsheim 1986).

Thinking of a school as an example may help. The boundaries may be defined by physical entities (e.g., the building and playground). Key components of the system include teachers, administrators, and the school board. The flow of resources into the system involves financial support and new instructional programs. Reports from teachers via parent conferences and press releases by administrators are examples of the flow of information out of the system. Finally, feedback comes in the form of assessments of student learning or program alignment with the state science standards; such feedback may confirm the system's current functioning or motivate changes such as new programs or professional development.

The science education system has many components that are often relatively simple to identify. The explicit theme of this book is expressed in the title: *Leadership by and for Science Teachers*. Science teachers and their teaching are essential components in the system and directly affected by the system's purpose, which is to facilitate and strengthen students' science knowledge and abilities. Other human components such as science coordinators and principals, and non-human components such as equipment and facilities, are important, but it is a good idea to identify interconnecting factors such as the flow of resources and information about the components that can affect the system's potential to accomplish its goals.

To evaluate your understanding of the principles discussed, answer the following questions with regard to the educational system you lead:

- What are the components of the system?

- What are the boundaries of the system?

- What information and resources flow into and out of the system (i.e., across the boundary)?

- What constitutes positive or negative feedback in the system?

- What behaviors best exemplify the system's purposes?

You might try answering these questions with colleagues in a professional discussion (i.e., a professional learning community). I also recommend involving the leadership team in the electronic version of the change game (The NETWORK Inc. 1999), which clarifies the system's dynamics and the need to think of the whole system in depth.

Systems, Standards, and Curricular Coherence

In her introduction to systems, which I discussed in the previous section, Donella Meadows (2008, p. 11) notes that a system is an interconnected set of elements that is coherently organized in a way that achieves something. One might note the implication that science educational systems are *coherently* organized. One guide for leadership is to provide some coherence within the system one is changing.

Good judgment recommends logical and orderly relationships among components of the educational system (Fullan 2001). This view is understandable. However, when one considers leadership in education, the general idea of coherence becomes more complex. Individual teachers' classrooms, colleagues' backgrounds and experiences, administrators' priorities, public perceptions of education—in each situation, the purposes, constraints, knowledge, values, and skills will influence the system's coherence. Coherence depends on the context within which one is establishing connections among the system components that contribute to the whole.

In an insightful essay, "Coherence in High School Science," F. James Rutherford (2000, p. 21) explains the term *coherence*:

> *In general, the notion of coherence itself is simple enough. It has to do with relationships. Things are coherent if their constituent parts connect to one another logically, historically, geographically, physically, mathematically, or in some other way to form a unified whole. Coherence calls for the whole of something to make good sense in the light of its parts, and the parts in light of the whole.*

Echoing the explanation of systems, leaders in science education face the challenge of clarifying the interconnected parts and the ways those parts interact to form a unified whole. This discussion limits the role of coherence to the essential features of the instructional core. My emphasis centers on those educational components that science educators will have to address in light of new science standards: curriculum materials, instructional strategies, and classroom assessments.

CHAPTER

23.

Conclusion

SEVERAL LEADERS USED THE WORD *journey* in their profiles. The word *journey* suggests a beginning, travels, and a destination. It also implies an understanding of the journey's purpose, one's means of travel, and the possibility of making decisions and confronting unexpected situations, both creative and challenging, along the way.

Recent reforms initiated as a result of new state standards for science education are neither the first nor the last calls for leadership journeys with the aim of improving science teaching and, subsequently, students' learning. Throughout the history of science education, teachers have been called on to set new goals, engage in curriculum reform, and respond to national needs for educated citizens as well as future employees in science, technology, engineering, mathematics, and related fields.

This perspective gives greater recognition to the role of science teachers and the leadership they can provide. The distribution of leadership must include science teachers, and if we want contemporary reforms to succeed, leadership must come from all of us. One or two or three leaders—even great ones—cannot achieve all of the tasks involved in reforming science education. The scale of reform is too large, the dispersal of power too wide, and the diversity of students, teachers, and schools too great.

One essential group that must be recognized for their leadership is science teachers. They are the leaders needed at the critical interface with students, yet, for too long, the larger community has not perceived them as leaders. It is well past time to change these perceptions.

As a science teacher, answering the question "How do I continue and expand as a leader?" will require some initial reflection. As the profiles of leadership in this book reveal, there are many situations and places in which one can assume a leadership role, whether large or small.

This book concluded with guides for your leadership. Although the chapter is wide ranging, I distilled the insights and recommendations from the contributed profiles into eight general features to help guide individuals as they assume leadership responsibilities (see Figure 23.1). The eight letters in the word *profiles* are used to outline the features that lead to exemplary leadership.

Figure 23.1. Essential Features of Leadership According to Contributors' Profiles

PURPOSE: An effective leader needs to be clear about his or her aims and express plans to accomplish those aims.

RELATIONSHIPS: Leaders must attend to personal relationships with those they are leading.

OPPORTUNITIES: Opportunities for leadership come in many forms. Leadership can begin by accepting an opportunity.

FACILITATION: A leader should be able to help a group work together to achieve a stated purpose. The leader might advocate for changes, provide resources, and work with individuals and groups, but the group as a whole achieves the goals. Facilitation implies collaborating and coaching, not controlling.

INTEGRITY: Leaders must act with personal and professional integrity. They demonstrate responsibility and support doing the right thing, using the best programs, and applying effective practices to enhance student learning. They also fulfill an obligation to maintain the integrity of science and related disciplines.

LEARNER: Leaders are learners who have a broad base of knowledge and abilities.

EQUITY: Effective leadership addresses inequalities in science education. The motto "All standards, all students" is an explicit statement of equity.

SYSTEMS: Leaders must understand the dynamics of system change and the contributions of resources, barriers, feedback, and various components.

Educators seldom discuss the features of leadership as synthesized from the contributors' profiles. I am sure no leader uses all of these guides all the time; I am sure that all science teacher leaders use some of these guides some of time. My point in this discussion is that if more science teachers apply these guides more often, we will see a steady evolution toward more effective leadership and increased constructive student learning—and, eventually, higher levels of scientific literacy among all citizens.

References

American Museum of Natural History (AMNH). 2018. Five tools and processes for translating the *NGSS* into instruction and classroom assessment. AMNH. *https://www.amnh.org/learn-teach/curriculum-collections /five-tools-and-processes-for-ngss*.

America's National Churchill Museum. n.d. Never give in, never, never, never, 1941. America's National Churchill Museum. *https://www.nationalchurchillmuseum.org/never-give-in-never-never-never.html*.

Anderson, R. D., and H. A. Pratt. 1995. *Local leadership for science education reform.* Dubuque, IA: Kendall/ Hunt.

Banathy, B. 1992. *A systems view of education: Concepts and principles for effective practice.* Englewood Cliffs, NJ: Educational Technology Publications.

Baxter, H. 2010. SIMPL: A framework for designing and facilitating professional development to change classroom practice. *Science Teacher Education* 59 (October): 13–28. *https://www.urbanadvantagenyc.org /wp-content/uploads/2018/08/SIMPL-Lauffer.pdf*.

Bennis, W., and B. Nanus. 1985. *Leaders: Strategies for taking charge.* New York: Harper and Row.

Berry, B. 2019. Teacher leadership: Prospects and promises. *The Kappan* 100 (7): 49–55.

Biological Sciences Curriculum Study (BSCS). 1999. Emerging and re-emerging infectious diseases, grades 9–12. NIH Curriculum Supplement Series. Colorado Springs, CO: BSCS.

Blake, J. A., K. W. Lee, T. J. Morris, and T. E. Elthon. 2007. Effects of turnip crinkle virus infection on the structure and function of mitochondria and expression of stress proteins in turnips. *Physiologia Plantarum* 129 (4): 698–706.

Branch, G. 2020. Anti-intellectualism and anti-evolution: Lessons from Hofstadter. *Phi Delta Kappan* 101 (7): 22–27.

Burns, J. M. 1978. *Leadership.* New York: Harper and Row.

Bybee, R. 1973. The teacher I like best: Perceptions of advantage, average, and disadvantaged science students. *School Science and Mathematics* 73 (5): 384–390.

Bybee, R. 1975. The ideal elementary science teacher: Perceptions of children, pre-service and in-service elementary science teachers. *School Science and Mathematics* 75 (3): 229–235.

Bybee, R. 1978. Science educators' perceptions of the ideal science teacher. *School Science and Mathematics* 78 (1): 13–22.

Bybee, R. 1993a. Leadership, responsibility, and reform in science education. *Science Educator* 2 (1): 1–9.

Bybee, R. 1993b. *Reforming science education: Social perspectives and personal reflections.* New York: Teachers College Press.

Bybee, R. 1994. Teaching science: Six guides for the journey. In *Exploring the place of exemplary science teaching,* ed. A. Haley-Oliphant, 63–82. Washington, DC: American Association for the Advancement of Science (AAAS).

Bybee, R. 2006. Leadership in science education for the 21st century. In *Teaching science in the 21st century*, ed. J. Rhoton and P. Shane, 147–162. Arlington, VA: NSTA Press.

Bybee, R. 2013. *Translating the NGSS for classroom instruction.* Arlington, VA: NSTA Press.

Bybee, R. 2015. *The BSCS 5E Instructional Model: Creating teachable moments.* Arlington, VA: NSTA Press.

Bybee, R. 2018. *STEM education now more than ever.* Arlington, VA: NSTA Press.

Bybee, R. 2020. *STEM, standards, and strategies for high-quality units.* Arlington, VA: NSTA Press.

Bybee, R., and C. Chopyak. 2017. *Instructional materials and implementation of* Next Generation Science Standards*: Demand, supply, and strategic opportunities.* New York: Carnegie Corporation.

Bybee, R., and S. Pruitt. 2017. *Perspectives on science education: A leadership seminar.* Arlington, VA: NSTA Press.

Bybee, R., W., J. A. Taylor, A. Gardner, P. VanScotter, J. Carlson Powell, A. Westbrook, and N. Landes. 2006. *The BSCS 5E Instructional Model: Origins and effectiveness.* Colorado Springs, CO: BSCS.

California Department of Education. 2014. INNOVATIVE: A blueprint for science, technology, engineering, and mathematics. Report by State Superintendent of Public Instruction Tom Torlakson's STEM Task Force, Professional Learning Support Division, Californians Dedicated to Education Foundation.

Calvino, I. 1988. *Six memos for the next millennium.* New York: Vintage.

Carter, S. 1996. *Integrity.* New York: Harper Collins.

Cheung, R., T. Reinhardt, E. Stone, and J. W. Little. 2018. Defending teacher leadership: A framework. *Kappan* 100 (3): 38–44.

Combs, A., D. Avila, and W. Parkey. 1978. *Helping relationships: Basic concepts for the helping profession.* Boston: Allyn and Bacon.

Combs, A., A. Misner, and K. Whitaker. 1999. *On becoming a school leader: A person-centered challenge.* Alexandria, VA: Association for Supervision and Curriculum Development (ASCD).

Covey, S. R., A. R. Merrill, and R. R. Merrill. 1995. *First things first.* New York: Simon and Schuster.

Cremin, L. 1965. *The genius of American education.* New York: Vintage Books.

Dewey, J. (1916) 1966. *Democracy and education.* New York: Macmillan. Reprint, New York: Free Press.

DiRanna, K., E. Osmundson, J. Topps, L. Barakos, M. Gearhart, K. Cerwin, D. Carnahan, and C. Strang. 2008. *Assessment-centered teaching: A reflective practice.* Thousand Oaks, CA: Corwin.

Duschl, R. A. 1988. Abandoning the scientific legacy of science education. *Science Education* 72 (1): 51–62.

Evans, R. 1996. *The human side of change: Reform, resistance, and the real-life problems of innovation.* San Francisco: Jossey-Bass.

Family of Vince Lombardi. n.d. Famous quotes by Vince Lombardi. *http://www.vincelombardi.com/quotes.html.*

Frankl, V. E. 1962. *Man's search for meaning: An introduction to logotherapy.* Boston: Beacon.

Fullan, M. 2001. *Leading in a culture of change.* San Francisco: Jossey Bass.

Fullan, M. 2010. *All systems go: The change imperative for whole system reform.* Thousand Oaks, CA: Corwin.

Fullan, M. 2013. *Motion leading in action.* Thousand Oaks, CA: Corwin.

Fullan, M., and M. Miles. 1993. Getting reform right: What works and what doesn't. *Phi Delta Kappan* 73 (10): 745–752.

Gamoran, A. 2019. Evidence lights the way. *Science* 365 (645): 843.

Georgia Department of Education. 2002. Released Georgia High School Graduation Test results.

Georgia Department of Education. 2010. Released Georgia High School Graduation Test results.

Gewertz, C. 2020. "The Art of Making Science Equitable." *Education Week*, March 3.

Guskey, T. 1986. Staff development and the process of teacher change. *Educational Researcher* 15 (5): 5–12.

Hall, G., and S. Hord. 2011. *Implementing change: Patterns, principles, and potholes.* 3rd ed. Boston: Allyn and Bacon.

Hall, G., and S. Hord. 2015. *Implementing change: Patterns, principles, and potholes.* 4th ed. Upper Saddle River, NJ: Pearson.

Harper's Bazaar. 2017. 21 of Maya Angelou's best quotes to inspire. Harper's Bazaar. *https://www.harpersbazaar .com/culture/features/a9874244/best-maya-angelou-quotes/.*

Heyman, D. (Producer), and M. Newell (Director). 2005. *Harry Potter and the Goblet of Fire.* United States: Warner Brothers.

Hofstadter, R. 1963. *Anti-intellectualism in American life.* New York: Alfred A. Knopf.

Jersild, A. 1955. *When teachers face themselves.* New York: Bureau of Publications, Teachers College, Columbia University.

Kierkegaard, S. 1951. *The sickness unto death.* Translated by W. Lowrie, Princeton, NJ: Princeton University Press.

King, Martin Luther, Jr. 1963. *Strength to love.* Boston: Beacon Hill Press.

Lawrence Hall of Science. *Disruptions in ecosystems: Ecosystem interactions, energy, & dynamics.* Berkeley, CA: Lawrence Hall of Science. *https://sepuplhs.org/middle/disruptions-in-ecosystems/.*

Little, J. W. 1998. Assessing the prospects for teacher leadership. In *Building a professional culture in schools,* ed. A. Lieberman, 79–106. New York: Teachers College Press.

Loucks-Horsley, S., P. W. Hewson, N. Love, and K. E. Stiles. 1998. *Designing professional development for teachers of science and mathematics.* Thousand Oaks, CA: Corwin.

Loucks-Horsley, S., R. Kapitan, M. O. Carlson, P. J. Kuerbis, R. C. Clark, G. M. Melle, T. P. Sachse, and E. Walton. 1990. *Elementary school science for the 90s.* Alexandria, VA, and Andover, MA: Association for Supervision and Curriculum Development and The NETWORK, Inc.

Loucks-Horsley, S., K. Stiles, S. Mundry, N. Love, and P. Hewson. 2010. *Designing professional development for teachers of science and mathematics.* 3rd ed. Thousand Oaks, CA: Corwin.

Maslow, A. 1970. *Motivation and personality.* New York: Harper and Row.

May, R. 1950. *The meaning of anxiety.* New York: Ronald Press.

May, R. 1975. *The courage to create.* New York: W. W. Norton.

McGowan, P., and J. Miller. 2001. Management vs. leadership. *The School Administrator* 58 (10): 32–34.

Meadows, D. 2008. *Thinking in systems.* White River Junction, VT: Chelsea Green Publishing.

References

Molière. 1670. *The middle class gentleman.* Project Gutenberg. *https://www.gutenberg.org/files/2992/2992 -h/2992-h.htm.*

National Academies of Science, Engineering, and Medicine. 2019. *Science and engineering for grades 6–12: Investigation and design at the center.* Washington, DC: National Academies Press.

National Research Council (NRC). 1996. *National science education standards.* Washington, DC: National Academies Press.

National Research Council (NRC). 2001. *Educating teachers of science, mathematics, and technology: New practices for the new millennium.* Washington, DC: National Academies Press.

National Research Council (NRC). 2012. *A framework for K–12 science education: Practices, crosscutting concepts, and core ideas.* Washington, DC: National Academies Press.

National Research Council (NRC). 2014a. *Developing assessments for the* Next Generation Science Standards. Washington, DC: National Academies Press.

National Research Council (NRC). 2014b. *Exploring opportunities for STEM teacher leadership.* Washington, DC: National Academies Press.

National Research Council (NRC). 2015. *Guide to implementing the* Next Generation Science Standards. Washington, DC: National Academies Press.

National Research Council (NRC). 2019. *Science and engineering for grades 6–12: Investigation and design at the center.* Washington, DC: National Academies Press.

Neem, J. 2020. Anti-intellectualism and education reform. *Phi Delta Kappan* 101 (7): 10–16.

The NETWORK, Inc. 1999. Making change happen! The NETWORK, Inc. *http://www.thenetworkinc.org /games/leadership-series/mch/.*

NextGenScience. 2021. *Toward* NGSS *design: EQuIP rubric for science detailed guidance.* WestEd. *https://www.nextgenscience.org/resources/toward-ngss-design-equip-rubric-science-detailed-guidance.*

NextGen TIME. n.d. Virtual NextGen TIME. *https://www.nextgentimepl.org/home.*

NGSS Lead States. 2013. *Next Generation Science Standards: For states, by states.* Washington, DC: National Academies Press. *www.nextgenscience.org/next-generation-science-standards.*

Nichols, T. 2017. *The death of expertise: The campaign against established knowledge and why it matters.* New York: Oxford University Press.

Olson, J. 2019. Teacher spotlight: Inside the high school classroom. *The Science Teacher* 81 (2): 56–58.

OpenSciEd. n.d. About OpenSciEd. OpenSciEd. *www.openscied.org/about.*

Phillips, D. T. 1992. *Lincoln on leadership.* New York: Warner Books.

Pratt, H. 1984. Science leadership at the local level: The bottom line. In *Redesigning science and technology education: 1984 yearbook of the National Science Teachers Association,* ed. R. Bybee, J. Carlson, and McCormack, 141–146. Washington, DC: NSTA.

Pratt, H. 2013. *The NSTA reader's guide to the* Next Generation Science Standards. Arlington, VA: NSTA Press.

Resnick, L. B. 1987. *Education and learning to think.* Washington, DC: National Academies Press.

Rhoton, J., ed. 2010. *Science education leadership: Best practices for the new century.* Arlington, VA: NSTA Press.

Rhoton, J., ed. 2018. *Preparing teachers for three-dimensional instruction.* Arlington, VA: NSTA Press.

Rhoton, J., and P. Bowers, eds. 1996. *Issues in science education.* Arlington, VA: NSTA Press.

Rhoton, J., and P. Shane, eds. 2006. *Teaching science in the 21st century.* Arlington, VA: NSTA Press.

Rodman, A. 2018. Learning together, learning on their own. *Educational Leadership* 73 (3): 12–18.

Rowe, P. G. 1992. *Design thinking.* Cambridge, MA: MIT Press.

Rowling, J. K. 1999. *Harry Potter and the chamber of secrets.* New York: Scholastic.

Rowling, J. K. 2007. *Harry Potter and the deathly hallows.* New York: Random House.

Rutherford, F. J. 2000. Coherence in high school science. In *Making sense of integrated science*, 21–29. Colorado Springs, CO: BSCS.

Rutherford, F. J., and A. Ahlgren. 1989. *Science for all Americans.* New York: Oxford University Press.

Schmidt, W., and C. McKnight. 1998. What can we really learn from TIMSS? *Science* 282: 1830–1831.

Schmidt, W., C. McKnight, R. Houang, H. Wang, D. Wiley, L. Cogan, and R. Wolfe. 2001. *Why schools matter: A cross-national comparison of curriculum and instruction.* San Francisco: Jossey-Bass.

Schmoker, M. 2018. *Focus: Elevating the essentials.* Alexandria, VA: Association for Supervision and Curriculum Development.

Schmoker, M. 2019a. Enhancing the power of less. *Educational Leadership* 76 (6): 24–29.

Schmoker, M. 2019b. Focusing on the essentials. *Educational Leadership* 76 (2): 30–35.

Short, J. 2006. *Analyzing standards-based science instructional materials: An opportunity for professional development.* New York: Columbia University.

Short, J., and S. Hirsh. 2023. *Transforming teaching through curriculum-based professional learning.* Thousand Oaks, CA: Corwin Press, Inc.

Short, J., and S. Hirsh. 2020. *The elements: Transforming teaching through curriculum-based professional learning.* New York: Carnegie Corporation of New York.

Smylie, M. A. 1997. Research on teacher leadership: Assessing the state of the art. In *International handbook of teachers and teaching*, ed. B. J. Biddle, 521–592. Dordrecht, the Netherlands: Kluwer.

Spector, M. 2020. "How Anthony Fauci Became America's Doctor." *New Yorker*, April 20.

Stiles, K., S. Mundry, and K. DiRanna. 2017. *Framework for leading* Next Generation Science Standards *implementation.* San Francisco: WestEd.

St. John, M., J. Hirabayashi, F. Helms, and P. Tambe. 2006. *The BSCS National Academy of Curriculum Leadership: Contributions and lessons learned.* Inverness, CA: Inverness Research.

Thorsheim, H. 1986. Systems thinking: The positive influence of STS on educational motivation. In *Science-Technology-Society NSTA Year Book 1985*, ed. R. Bybee, 236–242. Washington, DC: NSTA.

Tillich, P. 1952. *The courage to be.* New Haven, CT: Yale University Press.

Tyler, R. 1949. *Basic principles of curriculum and instruction.* Chicago: University of Chicago Press.

References

Weissglass, J. 1998. *Ripples of hope: Building relationships for educational change.* Santa Barbara, CA: University of California Center for Educational Change in Mathematics and Science.

Wenner, J., and T. Campbell. 2017. The theoretical and empirical basis of teacher leadership: A review of the literature. *Review of Educational Research* 87 (1): 134–171.

WestEd K–12 Alliance. 2007. Take AIM when choosing materials. *What's the Big Idea?* 6 (3): 1–2, 4. *https://k12alliance.org/newsletters/6.3_AIM%20part%201_Jan.07.pdf.*

Will, M. 2017. "Students Fare Better When Teachers Have a Say, Study Finds." *Education Week*, November 1.

Will, M. 2020. "Teacher of the Year Nominees Speak Out." *Education Week*, March 3.

York-Barr, J., and K. Duke. 2004. What do we know about teacher leadership? Findings from two decades of scholarship. *Review of Educational Research* 74 (3): 255–316.